International Poultry

THE SILKIE FOW

OTHER POULTRY BOOKS
BY Dr JOSEPH BATTY

Bantams —Concise Guide
The Silkie Fowl
Old English Game Bantams
Understanding Modern Game
(with James Bleazard)
Understanding Indian Game
(with Ken Hawkey)
Bantams & Small Poultry
Poultry Ailments
Sussex & Dorking Fowl
Sebright Bantams
Poultry Characteristics—Tails
Artificial Incubation & Rearing
Natural Incubation & Rearing
Domesticated Ducks & Geese
The Ancona Fowl
Concise Poultry Colour Guide
Japanese Long Tailed Fowl
Poultry Shows & Showing
Natural Poultry Keeping
Understanding Old English Game
The Orpington Fowl
(with Will Burdett)
The Barnevelder Fowl
Marsh Daisy Poultry
Sicilian Poultry Breeds
Rosecomb Bantams
Malay Fowl
Khaki Campbell Ducks & the Campbells of Uley
Brahma & Cochin Fowl
Rhode Island Red Fowl
Minorca Fowl
Sussex & Dorking Fowl

THE

SILKIE

FOWL

JOSEPH BATTY

Beech Publishing House
Station Yard
Elsted
Midhurst
West Sussex,

ISBN 1-85736-194-6

First published 1996
New Edition 2000

British Library Cataloguing-in-Publication Data
A catalogue record for this book is available from the British Library.

Beech Publishing House
Station Yard
Elsted Marsh
Midhurst
West Sussex GU29 OJT

CONTENTS

DEDICATED TO THE MEMORY OF MRS A CAMPBELL WHO WAS THE ORIGINATOR OF THE KHAKI CAMPBELL DUCK AND WAS A LEADING FIGURE IN THE IMPROVEMENT OF THE SILKIE FOWL

Opposite:
An advertisement which appeared in the ***Poultry World Annual,*** 1914

FOREWORD

This book was written after a number of requests were received from fanciers for a comprehensive work on the Silkie Fowl. Although one of the best known of breeds, there is still confusion on whether the breed is a bantam or large fowl. There are, in fact, two types, and a difference exists between the American and British approaches.

The emphasis is on the management of Silkies for the fanciers, who wish to show, and those who simply keep the birds as a hobby.

My thanks are given to Mrs S Flude, Secretary of the Silkie Club and to many of the members who assisted with information and photographs. Where possible their names have been shown below the photograph.

Extensive use has been made of out-of-copyright material in the now defunct (original) *Feathered World*. In addition, many books have been used for reference and these are listed in the Bibliography at the end of the book.

I hope this book will encourage others to keep this unique fowl, which has so much to offer in interest and utility.

J Batty

A COMPARISON OF BREEDS

<u>The Unique Nature of The Silkie</u>

The illustration opposite shows the unique feathering and structure of the Silkie Fowl.

Each of the breeds shown has normal, webbed feathers and sickle feathers in the tail, whereas the Silkie has soft down all over, including the mound-like tail.

Colours in Silkies are generally restricted to self-colours: White, Black, and Gold, and the so-called Partridge variety, which is really a modification of the Gold. The Blue shown is typical of the breed, but there is also a bearded variety and a bantam.

Silkie
(Blue)
Welsummer
Yokohama
(Black Red)
Wyandotte
(Gold Laced)
Modern Game
(Pile Cock & Hen;
Partridge Hen)
Old English Game
(Large Brown Reds;
USA Bantams)

Opposite

A *Detail* from a painting entitled ***The Poultry House*** by Carl Jutz the Elder. (1838 – 1916).

The significance lies in the presence of a cock and hens which are quite clearly of the Silkie–type, although not like the modern–day Silkie. They are larger and have longer legs, the latter with sparse feathering. They appear to be farmyard poultry, but are breeding to type in the sense that they are all with silkie feathers.

An American Silkie in 1871

From *The People's Practical Poultry Book,*

Wm M Lewis, 3rd Edition.

These rather well presented birds appeared in the above book which was published and reprinted many times.

SPECIAL NOTE ON ILLUSTRATIONS

Some of the illustrations are quite old and reproduction was difficult. Even modern photographs, especially of Blacks, created problems, and had to be shown as silhouettes.

Because it was necessary to show the shape and type, the silkiness of the feathers, or the feathered legs, some of these rather difficult illustrations have been included. Even Mrs A Campbell, wife of the founder of the Silkie Club, and pioneer poultry breeder, did not reproduce very well, because the original was quite small and had appeared in a Year Book on newsprint paper.

For all this we apologize and hope that all the illustrations, even some which are imperfect, will serve a useful purpose.

1

BACKGROUND

Courtesy: R B Swarbrick
A Champion

ESSENTIALS

1, Classification: Light Breed in UK and bantam in USA.

Interestingly, the Silkie is predominantly a broody which makes it unusual because light breeds are often non-sitters.

2. Origins: Asia; China, Japan, India or one of the other countries in that part of the world.

3. Egg Colour: Whitish, sometimes a creamy colour.

4. Principal Characteristics:

(a) Feathers which are fur-like, thus giving a rather fluffy covering.

(b) Constantly becoming broody; will lay a 'clutch' (around 18 eggs) and then will wish to hatch them.

(c) Make excellent mothers.

5. Nature: Easy to keep, of gentle disposition, only fair layers, not generally accepted as table birds (black flesh).

6. Attraction: Can be kept in a small space, easy to feed and breed with, a number of varieties, very quaint and attractive so ideal for someone wanting to keep show birds.

What Is The Silkie Fowl?

The Silkie is a breed of fowl belonging to the general group of domesticated poultry. In size it is very small and, as a result, in the USA it is regarded as a bantam, ie; a small fowl which is usually around 25 per cent of the size of a standard type of fowl. In the UK it is classified as a normal fowl, but the Silkie Club also recognize a bantam. The weights* for the standard (large) bird is 4lb (1.82 kilo) for male and 3lb.(1.36 kilo) for the female. It is classified as a Light breed and although many light breeds are not sitters this breed is the first selection when broodies are required. A clutch of eggs is laid and then the hen comes broody wanting to hatch the eggs.

In the USA the weights are slightly lower being 36 oz. for a cock and 32 oz. for a hen**. The birds not fully matured would weigh less. This is around the same size as other bantams; in fact, heavy breeds such as Cornish (Indian) Game would be heavier at 44 oz. for a cock and 36 oz. for a hen. Accordingly, the USA breeders are justified in calling the Silkie a bantam so long as they aim for a very small fowl.

In Britain the attitude appears to have been : ***The Bigger the Better***, within reason. James Lilburn (Feathered World Year Book, 1928):

> **Do not imagine that a Silkie is a bantam. I would always favour a larger bird, all things**

*Originally 3 and 2 lb respectively. Increased to present standards in 1991.
** The bantam in Britain is 22oz (600g) for the cock and 18 oz for the hen, which is not according to the normal formula of 25%.

being equal.

As a result of this approach many fine large birds have been produced. I well remember Silkies I kept more than 25 years ago which were truly miniature, but more recently the breed has become larger all round. Without the reference to 'bantam' the size is fairly immaterial, provided it is remembered that this is a small breed. *Extremely large* birds cannot be justified, but larger birds seem inevitable. The extra weight now part of the British standard allows for this fact.

In any event, the Silkie is an original breed and for that reason it can be argued that it is not the miniature of a large fowl and therefore not a bantam.* However, whilst not disagreeing with the argument, readers will know that there are natural bantams which have no large counterpart, or may have a large fowl, but the bantam considered to be a distinct breed from the large; this is seen with the Cochin and Pekin (bantam).

A Small Silkie of Long Ago.

***Dr A Campbell, Hon sec. Silkie Club, *The Feathered World Year Book*, 1930.**

Main Features

The main feature is undoubtedly the **plumage** which is made up of feathers which tend to be 'furry' rather like an animal. There seems little doubt that the name came from the nature of the plumage. Harrison Weir called it the "downy" or Silkie fowl. A quotation from him ***(Our Poultry)*** attributed to the naturalist Latham (1810) is as follows:

> **The feathers, having webs with no adhesion to one another, have the appearance of down or rather of hair, whence the name silken is also given to it. This fowl is white, and about the size of the common fowl; it is clothed on the legs, but only outwards till the beginning of the claws, and on the outward claw to the first joint of toe or nail.**

The reference to colour indicates that the early Silkies were white in colour. As will be shown later, there are as many as five colours in the USA and four in the UK. All are crested, but in the USA there is a variety of Bearded Silkies. These are now available in Britain.

There is no indication from where the beard came, but it was shown as a separate variety in the USA from 1874 in the White colour. None of the early British birds show this feature; neither did the standards mention the beard. The beard hides the wattles and the birds are very attractive, but do not add much to the overall impression of the breed.

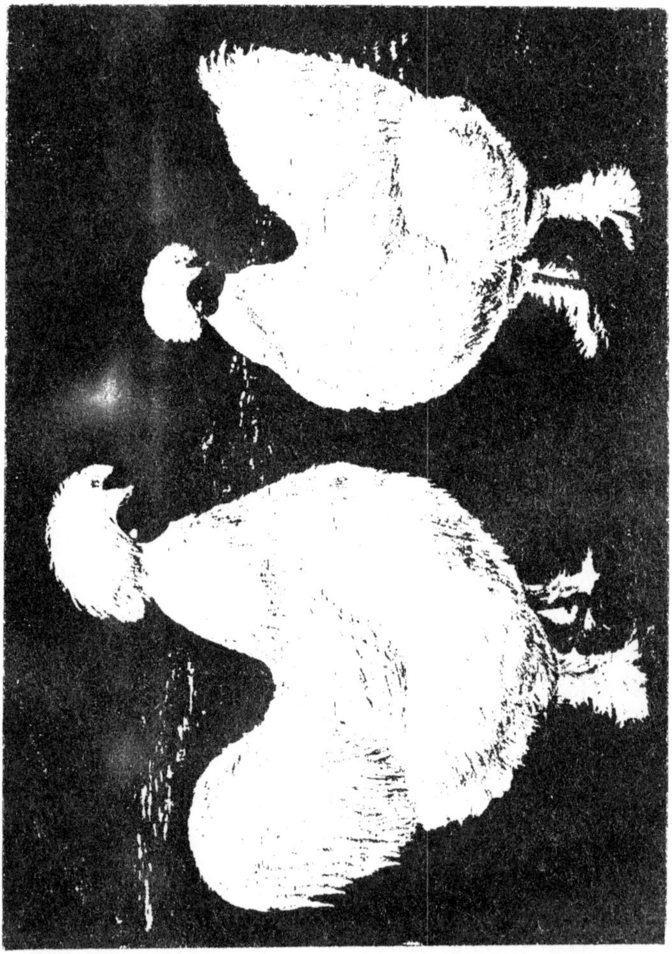

Silkies from a Painting by K Zander

Note the light feathering on middle and outer toes. The birds are showing more leg than modern birds

In 1776 there was a story in circulation in Brussels that there was a rabbit-fowl in existence which was the result of a cross between a rabbit and a fowl. The naturalist Buffon entered into correspondence on the subject and stated the 'culprit' was the Silkie.

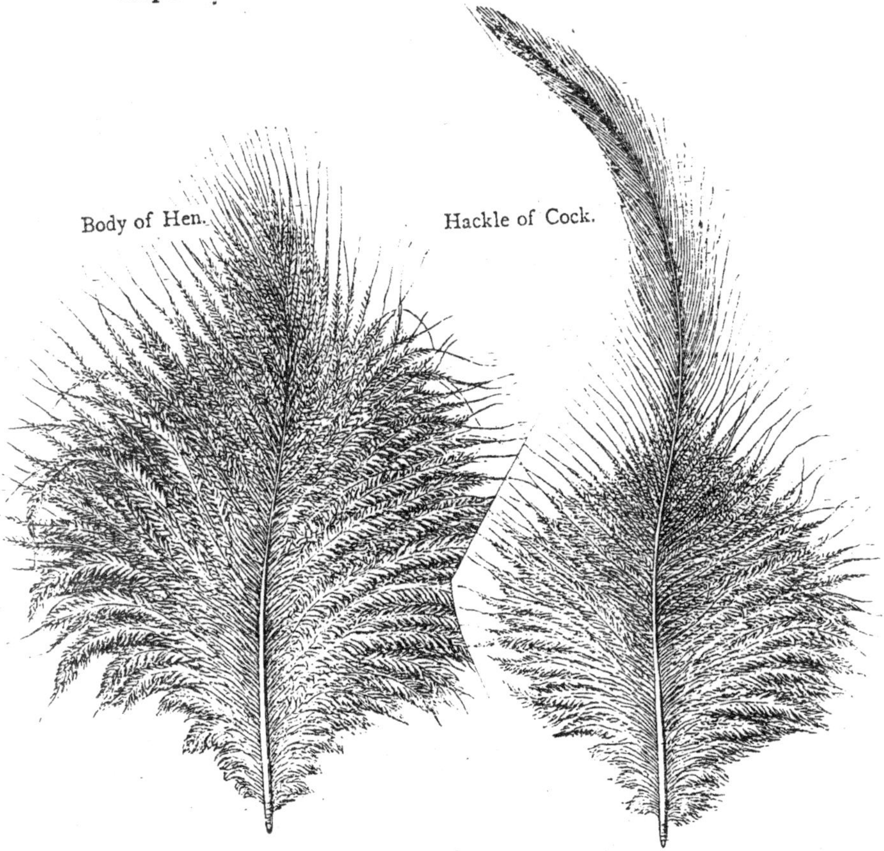

The Silkie Fowl's Feather

Any hint at the feathers not being downy must call for disqualification. As stated by a prominent fancier:

> **Without silk a Silkie is not a Silkie, and hard feathered birds should be discarded.**
> (James Lilburn)

The Fifth Toe

The Silkie has five toes, whereas the more usual is four. In that it is similar to the Dorking, Houdan and Faverolles. This condition is known as ***Polydactyly*** and is inherited.

The Crest & Beard

The breed may be

1. Non-bearded (with crest only)
2. Bearded with beard and crest

The crest should be full and round - and deformity should disqualify. Those with beards should have this feature quite full and extending backwards and outwards.

Feathered Legs

Legs should be feathered with soft feathers, but no vulture hocks (see illustration).

Link With The Flightless Birds

The reason for the Silkie being so different from other domesticated birds is baffling. The feathers, skin, bone, mulberry comb and turquoise ear lobes all make it very unusual.

There is very little resemblance to the Red Jungle Fowl from which most of our poultry descended. Yet the Silkie will interbreed with all other breeds of poultry, which means that the breed is part of the main poultry group.

The link with the flightless birds is interesting. The ***Ratites***, the group containing Ostrich, Emu, Fhea and Casowary have similar feathers and do not fly. Is the Silkie from the same evolutionary track as the Ratites? Certainly there are many similarities, although obviously the Silkie is a dwarf. The feathers are downy because they are simply used as a covering, but do not require the webs and adhesion for flight.

The early compilers of the Silkie *Standard* appears to have confused the Osprey bird with the term used in the millinery trade for display feathers of Egrets; these were referred to as "Ospreys" , but had no connection with the true Osprey. Certainly the wings are not "osprey plumaged" as suggested by the standard, which is also described as being a condition where some strands of the flights hang down, which is also not true of the Osprey.

Gene for Silkiness

For the technically minded the gene has been established as 'h' and is part of Chromosome 1 (Group III). Various geneticists have developed or modified the Linkage Map on chromosomes first developed by F B Hutt in 1936.* The silkiness is recessive to other forms of plumage, a fact recognized by Darwin and others.

Since absence of barbules on a feather is abnormal, it seems likely that silky feathers is due to some malfunctioning or malformation of the genetic structure.+ Even if this is the case, the characteristic has gone on for hundreds of years and therefore may now be regarded as an established feature.

---

*For those with a scientific mind reference should be made to *Genetics and Evolution of the Domestic Fowl*, Lewis Stevens, Cambridge, 1991, where the map can be studied.

+See *Modern Poultry Keeping*, G R Scott, London, 1927, where the author notes that Silkie-type plumage has appeared on other breeds, including Rhode Island Reds.

2

HISTORY OF THE SILKIES

THE ORIGINS

Many suggestions have been put forward to explain the origins of the Silkie Fowl. Some believe China, others Japan. Indeed, at one time the breed was known as the 'Japanese' Silkie.* Yet, as Wm Broomhead noted, that there had been no sighting of the breed in Japan. In the *Poultry Year Book* for 1927 the Editor is quite positive in asserting that :

> **This very distinctive breed originated in Japan, and is one of of the most remarkable specimens of the Japanese breeders' art.**

Undoubtedly, they came from Asia and besides the two countries mentioned Singapore and India have been given as possibilities.

The Silkie plumage has been bred on other breeds of poultry and this is noted elsewhere in this book. In the 1850s there were Buff and Partridge Silkie-type fowls, the latter being known as the 'Emu' fowl.* In addition, Langshans were bred with Silkie-type plumage.

Tracing the exact origins is quite difficult, especially when many decades are involved. In the next few pages a summary is given, showing the views of various writers throughout the ages. In modern times, the Silkie Club has done much to further the breed, being founded in 1898.

******Poultry Breeding & Management*, Wm W Broomhead, London, nd. He edited the *Poultry Club Standards* from 1910, and was responsible for many editions (subsequently this became the *British Poultry Standards*).**

EARLY REFERENCES*

The Silkie has been known for a considerable period. According to various authorities they appear to rank very high on a list of poultry by date of origin. The great explorer Marco Polo (1254 - 1324) recorded seeing the breed in China. The book describing his travels was written in French and translated into English, the best being by Sir Henry Yule in 1871. There is reference to the domesticated fowl with hair instead of feathers.

Other references usually cited are:

1. Konrad Gesner

2. Aldrovandi of Bologna

3. Ray translated by Willoughby

This author regarded the existence of a hairy bird as a figment of the imagination.

4. W B Tegetmeier (Victorian)

Tegetmeier, a friend of Darwin's knew the Silkie very well, but felt it was of very limited value. He stated (*Poultry for Table and Market*) when discussing profitable breeds:

> **Such are the black-skinned, black-boned Silky fowls, characterized by a white downy plumage, which contrasts strongly with the black combs and skin. These birds are not without their value as furnishing the best of all possible mothers for rearing pheasants and other small game birds;but as farmer's fowls they are**

*The coverage is for Silkies only, but it should be noted that there were early references to domesticated poultry from around 600 BC. In China birds were kept about 3,500 years ago and in India about 3,400. In Britain they date from the time of the Roman invasion (c. 50 AD).

infinitely worse than useless, as the introduction of a single bird into the farmyard will lead to the production of cross breeds, which, although perhaps neither black in the skin, nor silky in the plumage, have black bones and layers of black tissue beneath the muscle, rendering them unsuitable as table poultry.

He went on to state that the breed was good to eat, but the housewife, seeing the black flesh, would hesitate to eat it, believing that the bird was diseased.

5. Harrison Weir

Harrison Weir in *Our Poultry* (1903) quotes Latham (1810) who described the Silkie in detail.

He also states they were around before 1850 (See under 7 below). About the same time a Mr Cross showed a black "Silkie" at the Surrey Zoological Gardens, but this may not have been a true Silkie.

6. Buffon, G-L Leclere (1707 - 1788)

Buffon the French naturalist acknowledged the existence of the Silkie, but became involved in an attempted "fake" which was said to be a rabbit fowl, produced by crossing a rabbit and a common fowl. Buffon stated that the match was a physical impossibility.

7. Sir Edward Brown in *Poultry Breeding & Production,* 1929.

This author and poultry historian confirms Gesner (*History of Birds*) and Aldrovandi (1645) , but also states that the first specimens to reach this country was in the year 1850. There is no way of disproving this statement except by referring to other authors. Also in view of the antiquity of the breed this seems rather late.

Entwisle (*Bantams*) refers to the breed as if well established for a long time and Harrison Weir (ibid) refers to a writer Nolan who had kept the White Silkies many years prior to 1850 the year when he was writing. These birds had been imported from Sumatra and were used for hatching pheasant eggs.

Some confusion appears to exist between a "Negro Fowl" found in South America and other parts of the World and the true Silkie. Harrison Weir certainly believed the two were different, but the evidence is not conclusive. They may have been crossed with Silkies to give the dark skin. Another possibility is that some of the earlier species did not have the full coverage of Silkie feathers -- certainly some of the earlier illustrations were not as "woolly" as those seen today.

MODERN SILKIES

The start of the modern Silkies can be dated from around the time of the early activities of Mrs A Campbell of Uley in Gloucestershire (around 1900) who became well known for the breed of duck known as Khaki Campbell. She was also the breeder and exhibitor of Dorkings, and used Silkies as broodies, as well as winning many prizes with them.

In the 1920 issue of the year book, ***Poultry World Annual*** she wrote an interesting article on the breed at that date. One of her comments was that her Silkies were good layers of 2 oz size eggs and were in full lay early in January.

Mrs A Campbell: Poultry Pioneer & Breeder

Harrison Weir gives photographs of Mrs Campbell's Silkies in *Our Poultry* published in 1903. She was still showing into the 1920s and in *The Feathered World Year Book* for that year she comments on the fact that there were no shows during the last four years (due to the war), but then re-started in 1919. She took first prize with a hen and third with a cock and the hen "had wonderful silk".

The excellent ageing of Silkies was also commented on by Mrs Campbell. The cock was seven years of age and was still going strong. Mrs Campbell concluded that "Few breeds can be said to actually improve with age as the Silkie does."

At this time the Blacks and Partridges were being bred which gives some idea of the dating of the two varieties.

The Show Career

The breed was first classified at the Crystal Palace Show in 1872 and has been shown ever since. The numbers have never been very large, but it has an active club. Moreover, in assessing popularity it must be remembered that many breeders keep Silkies only as broodies for rearing other bantams or pheasants.

Numbers of entries at the National and International shows, starting in 1952 were 18 rising through to 72 in 1964 and then a decline to 22 in 1974. In the 1980s and onwards the numbers have once more soared so that a turnout of

Mrs Campbell's Silkie

Photographed at the beginning of the 20th century.

around 100 has been achieved by the joint efforts of the Silkie Club and the Scottish Silkie Club. Amazing progress for a so-called minority breed.

Today the breed seems to be extremely popular and there seems no reason why this should not continue. Whites continue to be the most popular and other colours are very much in evidence, although they do not appear to reach the standard in silkiness of the Whites. The Partridge colour, although a long time in existence, still strive for perfection; some strains are good, others are simply what appear to be poor quality Goldens.

Black Silkie Hen

Mrs Campbell; lst & Cup, Crystal Palace, 1912

Winners from the 1930s

Hen possibly vulture hocked. White Cock hard in tail feathers.

3

THE

STANDARD

REQUIREMENTS

Standard Features of the Silkie

SOURCES OF INFORMATION

The *Standards* are to be found in various works of reference; for example:

1. **Silkie Club in Britain with a separate one for Scotland.**
2. **American Standards of Perfection.**
3. **American Bantam Association.**
4. **British Poultry Standards.**

Standards are detailed specifications of the ideal or desirable bird for a particular breed and its varieties. These have been compiled by Club members, often over a long period, although some standards are very similar to those compiled by earlier fanciers as early as Victorian times.

Usually points are awarded for particular features and when birds are judged there should be a comparison made between the ideal feature; eg, Tail, and the actual tail, awarding points on the basis of how far a particular bird falls short of the top level of perfection.

Once this has been done for each exhibit in a class it becomes possible to make a comparison and to list the entries in order of merit. This listing then allows awards to be made.

If an exhibitor is to succeed then it will be necessary to know and understand the standard. The understanding part is vital because the process of judging is quite complex, and only those birds with minimum faults should be selected for showing.

An Early Standard

An early standard compiled by William Flamank Entwisle and fellow fanciers was available at the latter part of the nineteenth century. This is very basic, but does allow the main features to be seen. Later standards have been expanded and more detail given on some aspects to clarify points of difficulty.

Crest: Of the cock, pointing backwards; of the hen globular.
Comb: Strawberry, or half walnut, small and placed between the beak and the crest.
Beak: Short and curved.
Eyes: Bright but not very prominent.
Ear lobes: Of medium size.
Wattles: Of the cock, rather long and pendant; the hen's being small and round.
Head: Moderate in size.
Neck: Medium in length; the hackle very abundant, and flowing well on to the saddle.
Back: Very short, the ***Saddle*** of the cock being very broad, and rising almost from the neck hackle, and falling again at the tail, the ***Cushion*** of the hen being well rounded.
Tail: Very small and short.
Tail Coverts: Extremely soft and abundant.
Wings: Short and carried closely.*

*** See page 25 for footnote**

Breast: Broad, full and round.
Thighs: Short and well covered with fluff.
Legs and Feet: As short and as profusely feathered as possible, ***with vulture hocks***;* the toes, five in number, also being feathered, and well spread out.
Plumage: Soft, silky hairs and down instead of feathers, as ample and fine in quality as possible.
General Shape and Carriage: Compact, rather strutting, and active.
Weight: The cocks should not be more than 15 oz. nor the hens more than 12 oz.
Note: This was the reference to the early bantams weights which are now about 6 oz. more.

Source: *Bantams*, Entwisle, ibid.

This ***standard*** covers the essentials and is not in the form of either the British or the American standards. However, it will serve as a basis for explaining the requirements of the ideal bird and where there are deviations, one from the other, these will be explained.

Footnote p 24: **The reference to Osprey feathering is not included (as in the modern standard). The inclusion gives the wrong impression because the assumption is that the reference is to the Osprey bird, which is incorrect.'Osprey plumes' are probably a special type of plumage from an ostrich or similar bird.**

*** My italics. This is now a serious defect so should not be present.**

POINTS ON THE STANDARD REQUIREMENTS

If reference is made to the Entwisle standard it will be seen that all the vital requirements are covered. In this section the analysis is taken further and specific points are expanded.

Crest

The crest adorns the head.

In the **cock** it should be full with a form of 'tassel' or 'streamers' of about 1.5 inches (3.75cm) long. The crest should not obtrude or obstruct the face in any way or cover the eyes.

The **hen's** crest is round and smaller than the cock's. It is usually described as a 'Powder Puff' which does not split down the middle at the comb. The USA standard specifies a globular comb which is large and upright.

In both sexes the crest should be soft in texture and should be well balanced on the head. Very large crests with ugly combs should not be tolerated.

Note: The **beard** when it appears should be thick and full. It should cover the lower part of the face completely. The ***Bearded*** is a separate variety.

Comb

This is sometimes described as Walnut or Strawberry, and at other times Pea, but none of these is really correct. It is also referred to as a ***Cushion*** comb, which is possibly more accurate.

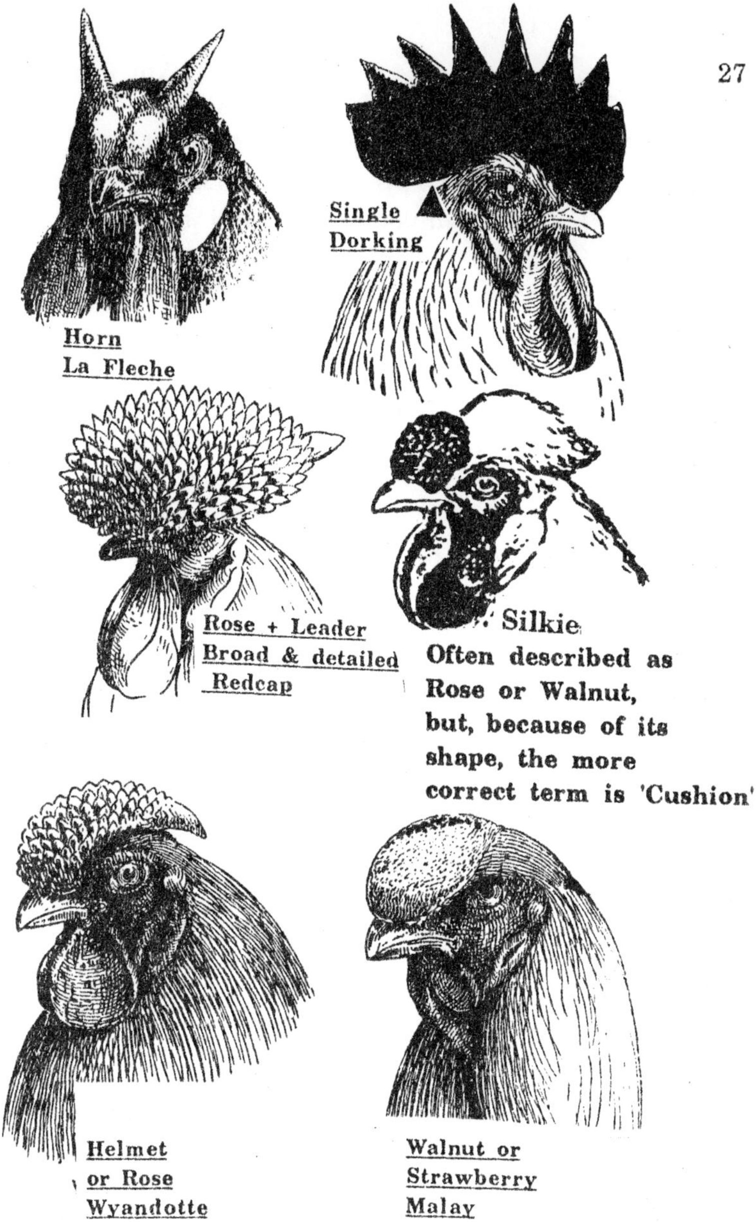

Diverse Types of Comb

The comb of the cock is very prominent and sits, some may suggest *perches,* at the front of the crest. It is quite prominent and circular. Indentations or ridges transverse the comb, and these are circular and there should be one main indentation across the middle.

The comb of the hen is quite small, very much hidden by the round crest. It does not display the same width or height of the cock's which is understandable between male and female, but the comb of the cock tends to be **rather exaggerated,** even allowing for the sexual differences.

Many combs on cocks are too lumpy and large, often detracting from the beauty of the bird. Colour is also a problem. Mulberry of a sound dark colour should be the aim; in fact, some combs are patchy or mottled with a dark red appearing on the mulberry; others are all over red, especially on older birds. Yet there is no shortage of stock with the correct colour.

The face should also be mulberry and the ear lobes should be a Turquoise colour. The latter does tend to vary from a clear colour to different shades and some birds even have white ear lobes which is unacceptable.

In the past some have suggested a lighter comb should be allowed for the 'Coloured' varieties, but since the very dark face and comb are found in the whites, this argument seems unrealistic. This mulberry colour is quite distinctive to the breed and therefore should be retained.

An American Bearded Silkie Hen

Champion, W J Rider

Photo courtesy Irvin Holmes

Beak

The beak should be fairly stout. The colour matches the face and is generally grey or slate at the base appearing horn or whitish at the tip.The British standard states 'mulberry' and the USA gives 'leaden blue'. In reality the colour does tend to be lighter at the tip of the beak as stated. Green on the beak is a serious fault.

Eyes

These should be quite bright and black in colour. The British standard states not too prominent, but the USA standard expects them to be prominent. If 'prominent' means standing out it would appear the the American standard is wrong; on the other hand, if it means 'Bold' then this is acceptable.

Ear Lobes

The colour which is the most attractive is Turquoise and this should be the aim. At present too much licence is given by judges. They should be of a reasonable size -- obvious, but not standing out.

Wattles

The wattles should be the same colour as the comb and the face. This seems an obvious statement, but many cocks have red wattles and a very dark comb.; ie, not matching, which looks strange and should be discouraged.

In the bearded variety the wattles should be diminutive.

The modern *standards* state that the wattles should *not* be pendant, but these seem out of touch with modern non-bearded cocks. In fact, many examined by the author do have quite long wattles. However, the hen's comb tends to be small and half circular.

Head

The head is relatively small, although with the embellishments of comb, crest, and, where appropriate, beard, it appears bigger.

If the overall head looks small and pinched the bird never looks right; the head should set off the rest. However, avoid excessive size in all aspects of the head and its embellishments or the bird, usually a cock, can look quite grotesque.

The head of the Silkie, including the comb is given an allowance of 30 points, which is a very important part of the bird's makeup.

Neck

The neck should be thick at the base with a thick hackle which falls on to the shoulders. The base should meet the base of the cushion thus giving the appearance of very little or no back.

Back, Saddle and Cushion

The back should be broad and short, a condition often described as stout, stocky or cobby. The hackle and cushion fit at each end and hide the actual shape. Only when a bird is handled can the true shape be felt. Younger birds may not be fully developed. In recent years some of the Blues, winning prizes have lacked cushion and have had high backs, indicating foreign blood, and this must not be accepted by judges, because this does not allow for the essential shape, including short back.

Example of Poland Crest
This is quite wrong for the Silkie; Too large and overhanging

Tail

Broad and ***short*** are the two key words with the end finishing in a curve with no hint of hard sickle feathers. It should be well rounded and fit into the overall conformation of tail and saddle.

The height in relation to the head and hackle is what has been termed the 'Gravy bowl' effect, the neck the spout, and the tail the handle.

This description of the tail applies to all colours, but many 'coloured', -- Blues, Golden and Partridge -- tend to have higher tails with a hint of 'hardness' at the top of the tail.

Historically, there has always been some difficulty even with Blacks. Mrs Cambell* (ibid) noted that there were problems:

> **The difficulty seems type in colour; this will take time. At present many are too leggy. The Blacks stick very much. I have gone so far with them and cannot get any further. It seems impossible to get more silk on them, but if one is keen enough to keep on success generally comes in the end.**

Fortunately, the Blacks now seem to be up to standard, but many Golden and Partridge are not up to standard.

In achieving the 'ideal' tail with no sign of rigidity there is a danger of reducing the size of the tail and losing essential symmetry. This should be avoided because the breed must not finish up as a half-rumpless fowl!

*** *Feathered World Yearbook,* 1928. At that time Mrs Campbell had been breeding them for 30 years!**

A Sultan Displaying 'Vulture Hocks'

These are stiff feathers growing downwards at the hock in a V-shape. They should not be on a Silkie.

Breast

As noted, a Silkie must be **full and rounded,** so the breast should fit this description. It should be prominent, curved and well forward, but not like a fantail pigeon or even an Old English Game bantam cock, because the angle of the breast indicates character. The Silkie cock is a gentle bird, proud but not showing the fact too much in his demeanour. The overall body carriage is horizontal.

Thighs

These should be well covered in fluff. At one time the legs were very clearly seen, but now the body fluff in a good female specimen almost reaches the ground so only the side of the foot is visible.

Cocks are taller and tend to show more leg, but not as long as previously. It has been suggested that the movement is towards a Pekin-type Silkie, but this seems quite wrong and unacceptable because it would change the character of the breed.

Legs and Feet

Initially it seems that vulture hocks were permitted, but this is no longer the case. In fact, except on certain breeds, such as the Sultan, the vulture hock is frowned upon. The Brahmas are seen with vulture hocks, but are best without because the leg feathers are never the same when vulture hocks are present.

Shanks should be short. Indeed, the tendency is for the lower body feathers (often called the *Fluff*) to cover the thighs and often only the inner toes are visible. The feathers on the outside of the legs complete this hiding process so these can also be seen underneath the lower body feathers. These feathers should not be too strong in growth because this would be out of character; a Silkie should not have any hard feathers at all. Nevertheless, the soft, silky feathers should be plentiful and not broken or damaged in any way.

Keeping the leg and foot feathers in sound condition can be a serious problem. If the birds are kept in wet or muddy conditions the feathers become damaged – they get broken, dirty and worn. In addition, unless the legs receive regular attention they are subject to Scaley Leg a condition which results in the legs becoming quite disfigured. The only safe way is to keep the birds on a bedding of clean, dry shavings and wash the legs at regular intervals. If there is any sign of roughness on the scales it is advisable to rub on sulphur ointment or a mixture of sulphur and medicinal tar obtainable from the chemist. Some fanciers suggest dipping the foot and shank in paraffin (or paraffin and linseed oil), but this can be dangerous because if any gets on to the skin it burns and can do considerable damage as well as being painful to the bird affected.

The Silkie has five toes and the male is

Scaley Leg

The Five Toes

The Leg

Shown on clean legged birds because it would be difficult to illustrate clearly on the feathered leg. A Silkie should have lightly feathered legs and on outer and middle toe (See p 6 for illustration).

spurred, the spur being medium in size and is placed just above the fifth (back) toe. The toes normally found on poultry – three at the front – should be evenly spread, and the back should be well down. It is the middle and outer toes which are feathered. If the middle toe is without feathers this should be regarded as a serious fault, but winning specimens are seen with this fault so some latitude is allowed by the judges.

The stance of the Silkie requires the legs to be set apart and straight when looked at from the front. Fanciers who are puzzled by this stipulation should compare with Old English Game where the legs must not be straight, but should follow the line of the body so the cock is 'close heeled', referring to the position of the spurs.

Plumage

The distinctive characteristic of the Silkie is the fact that it must be extremely soft in the feather so that the feathers resemble a silky fur. The older writers used to refer to birds having 'lots of silk' ; Mrs Campbell, writing of a winning bird in 1920 stated:

"My 1st prize hen has wonderful silk. It reached her feet, no leg at all showing".

Obviously some soft quills cannot be avoided on the tail and in the wings, but there must be no rigidity. Furthermore, the main feathers should be what has been termed 'shredded'. The normal webs of the feathers are absent so there is no adhesion of the barbs.

The Silkie

Hard Feathers Showing (a Fault)

Showing Great Profusion of Silk

Comparing the Types

General Shape & Carriage

The overall impression should be of a fluffy, rounded bird, compact and cobby, yet quite active. Although there is a certain amount of strutting the breed is usually of gentle disposition so birds can be handled and become quite tame. This is to be expected of a breed which is so renowned for hatching eggs and rearing chicks.

Weights

The weights are quite small being 1.81 kilos (4lb.) for male and 1.36 kg. (3lb.) for female for the standard bred British fowl. There is a tendency for larger birds to be produced and shown. Bantams are around half the size, although these are not generally available and, some would argue, seem to serve little purpose with such a small breed. The fact remains that the Silkie Club does give standard weights for bantams - 600g (22oz.) and 500g (18oz.) - so the interest does exist.

In the USA , where the breed is classified as a bantam the cock is 36oz. (cockerel 32oz.) and the hen 32oz. (pullet 28oz.).

4

COLOURS

AVAILABLE

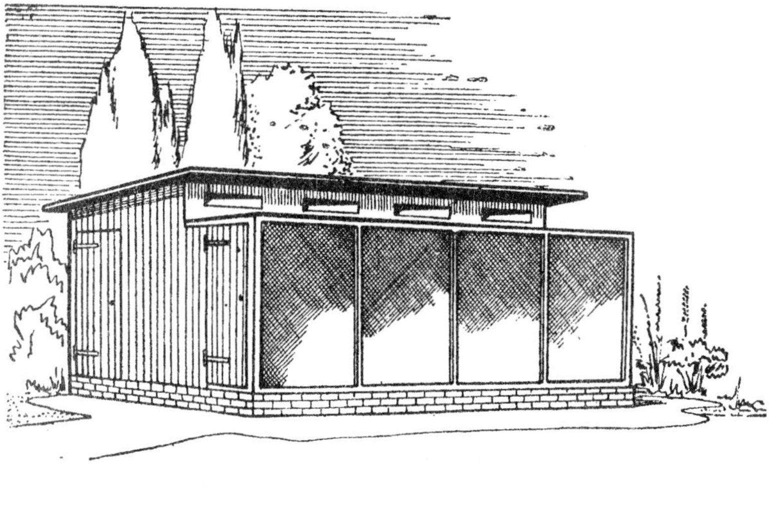

Colour/Variety Recognition

The colours which were recognized from early times were White and Black. Some of the very early references; eg, Marco Polo, stated that the silkie-type fowl was black. However, It would appear that the first colour to be domesticated in its present day form is the White.

Table

White Black Golden	**Shown from 1874 or earlier UK *standards* in 1925**
Partridge	***Standard* 1993 in UK**

Bantams

In the USA the standards have been recognized for many years. This is because Silkies are recognized only as bantams. In the UK the bantam standard was not issued until 1992.

DEVELOPMENT OF COLOURS

The natural colour for poultry is Black/red which means a cock with black breast and red hackles and upper body. The matching hen is partridge or a variation of the colour. Only when a melanistic black is introduced is this pattern disturbed because, as evidenced by the Black Breasted Dark Reds, the females can then be purple black with a black hackle ticked with burgundy.

Over a long period a number of basic colours have evolved; they are Black, White, Blue, Buff, Red and Yellow. Accordingly, the Silkie has a relatively straight forward colour pattern because three of the main colours are in the list. The difficult one is the Partridge which combines a number of colours and distinct markings. For this to be established it is clear that other breeds had to be used and, as a result, the feathering has suffered. Also the exact feather patterns on the conventional Partridge colouring cannot be produced on what effectively is fur, not feathers. If the rabbit is considered there are definite colour patterns, but these tend to be in 'blocks' and so it must be with the Silkie; there are specific colours, but it seems pointless to stipulate lacing on feathers when the feather is not supposed to exist.

Tails tend to be too high and too stiff. This also applies to Goldens and Blacks, but possibly not to the same degree. As noted, black is allowed on the tail.

In recent years more research has been carried out on colour. In particular it has been noted that the colour seen is dependent on the reaction of the light on the subject matter. In addition, the colour is affected by the pigments present in the feathers. These are known as 'melanins' and 'caratenoids'.

The caratenoids are an influence on the orange or yellow colours (Buffs). The basic yellow colouring is referred to as lipochrome and this forms the background for the further colouring. However, the melanins tend to be more important. They have a great influence on colour and may be looked upon as the dark pigments.

The melanins are produced by the melanocytes which exist on the feathers. The main types are the Black (***eumelanin***), and the Red (***phaeomelanin***) and it is the production of these shades which determine the final colour.

Whether a colour is Dominant or Recessive is also of great importance, but it should also be appreciated that colours may react differently because of the make-up of the remaining colours. The existence of a colour may prevent a colour being shown at its full strength. This is a complex subject which may be studied from a specialized book on the subject. In breeding it is important to appreciate that the colour mixing is not always as might appear.

RANGE OF COLOURS

As noted earlier, there are a number of colours which have been in existence for many years. These are as follows:

1. **White**
2. **Black**
3. **Golden or Buff**
4. **Blue or Grey**
5. **Partridge**

White

The Whites are the original colour, although earlier references do mention a Black Silkie-type bird. The overall colour should be as brilliant white as possible. This can be difficult to achieve because the nature of the feathers, not being very glossy, tends to be a dullish or soft white.

Serious faults in the white tend to be the existence of a cream or yellowish tinge, often brought on by exposure to the sun. Constant feeding of maize may also cause a creamy tinge to appear, so try to limit the feeding of this corn.

Breeding of white from sound whites is quite simple. Therefore selection is vital, avoiding any birds with colour markings or tinges. Occasionally, the use of a Black cockerel is said to be beneficial, thus improving the white by selection from the new stock. However, this practice is not a general recommendation.

This colour is often called 'snow white', and

to achieve the brilliance washing, followed by a blue rinse, is often advocated.

Other colour requirements are for the eyes to be black and the legs, feet and beak to be a slatey blue or grey lead colour. The Black tends to have a darker colour than, say, the white.

Nails should be a bluey-slate/white colour. Some variation may be found, the lighter birds having a tip of white on the beak (blending into the darker colour), and a horn colour on the toe nails.

Champion White Silkie Cock

Acknowledged as possibly the best example in recent times; note the 'ragged' feathers and abundant silk. The tail is probably as short as is feasible or the overall shape would be spoilt. (W Heaton).

Blacks

Being a self colour the Black should be that colour and no other. The male Silkie is required to have a greenish sheen to the feathers.

Some leniency is allowed in young birds, possibly lacing in the hackle, but this should not be encouraged so any with a sign of variation would be placed below the perfectly coloured birds.

In Silkies the colour black appears to be dominant so mating sound black male and female together should give good results. This is not the case with many other breeds where blacks produce blues and splashes. However, these colours are Blue/blacks and not Green/blacks and this is probably the reason for the difference in results. In Andalusians and Old English Game this mixing of the colours is the normal expectation. However, with the melanistic Blacks with purple, green and very glossy plumage I have found that such birds breed true.

On this assumption we would be led to believe that the Blue Silkies are not related to the Blacks, but, in fact, some strains are based on a cross between Blacks and Whites with Splashes also being involved*. However, this is a long process and results in many rejects before achieving stability. Accordingly, this may not be the best way of producing or improving Blues.

***See the account by Fred Jeffrey in *Silkies* (American Bantam Association) . This was on the experience of a Mr Polston.**

Fortunately, there are now many Blues in existence so the need for crossing does not appear to be necessary, any new blood coming from existing stock. When new stock is acquired make sure (if possible) that the birds are bred from blue on both sides or a multitude of colours will be bred.

Silhouette

Two Black Silkie Females

Note the profusion of silk and nicely rounded tail as well as density of Black: too many Blacks tend to be patchy without proper density of colour. (T Penny)

Golden Or Buff

Buff has always been a popular colour for poultry fanciers. It is seen in Orpingtons, Sussex, Wyandottes, Old English Game (Gingers) and many others. The main problem is to get the colour deep and even, and to prevent fading. Pale yellow resembles a 'washed out' buff and must be avoided. The colour becomes paler with age so it is wise to breed from stock with sound, deep colours.

A number of breeders in the USA have recorded how they 'produced' the Golden Silkie and for details the reader is referred to *Bantam Breeding & Genetics* , Fred P Jeffrey. However, it does appear that the variety did originate in Britain, no doubt from a cross with a small breed such as the Buff Pekin and then by careful selection, the standard type was established.

Some black will appear in the tail and this is permissible. Many Golden cocks have a dark colour across the shoulders and the sickle feathers are a shiny green-black with webbing at the top. These features are signs of a mixed ancestry and it is hoped that they will be bred out so they are as silky as the pure bred Whites. Females tend to be more evenly coloured in a golden, deep fawn. Even so specimens do show lighter patches on the 'trouser' part of the thighs.

Now a deep coloured is stabilized the reproduction of the colour presents no serious problems, but avoid hard feathering on breeding stock.

Partridge Colour

With Silkies there is bound to be problems in achieving a true partridge colour. This is the case because the male should be a Black Red and the female a brown bird with stipplings and pencillings to give the appearance of a partridge. The very nature of the plumage makes exact colour patterns difficult to achieve; how can pencilling be formed when the feathers are not made up of an even, flat surface.

The colour therefore should be regarded as an overall impression rather than a precise description as expected from more conventional breeds. Generally a Partridge hen would have the following features:

> **Hackles golden streaked with black; Body partridge, without rust or reddness (feathers finely pencilled); Tail black.**
>
> **The cock would have a red hackle and shoulders, with a black breast, thighs and tail.**

Many Partridge Silkies tend to be Goldens with a mingling of darker feathers. Moreover, they lack the precise markings of the normal partridge description found in Old English Game and many other breeds.

The Club *standard* states that the orange hackle of the male should have a black stripe in the centre. The American Bantam Association stipulates that the colour of hackle should be black with lacing of red and the saddle should be a rich red of a medium shade.

The ABA stipulate that the upper back should give the overall impression - when viewed from a distance - of rich red. In other words, the detailed red edging around the black feathers should give the red effect, although this is mingled with black.

The female is lemon on neck and breast with black stripes. The body should be a partridge colour with black barring. The wings tend to appear quite dark and it is this feature which distinguishes this variety from the Golden which should be a golden buff. The tail has some black in both colours.

Special Note: At the time of writing it appears that the Partridge colour is struggling to achieve the colour stated in the standards. Although around for a considerable period the difficulty of maintaining a perfect colour for this variety appears to be a formidable tasks. Yet it is an attractive colour which attempts to follow the wild plumage pattern of the Red Jungle Fowl.
Possibly the difficulty in breeding the colour consistently has led to the delay in standardization. As a variety it was in existence from the 1920s and possibly earlier, but only in 1993 was the Partridge standard added, and, even now, there are still difficulties.

Blue or Grey

Blues have always fascinated fanciers. In Silkies they should be a slate colour, evenly spread over the body. In males a darker shade is usually present on hackles, saddle and tail.

Some variation in shades will be found, but the more blue they become the more is the tendency to get a rather washed out appearance. An intermediate level is desirable, but to maintain this state it will be necessary to mate the lighter hens with a dark cock.

As noted earlier under the *Black* section, Blues may be related to the Blacks, but this is not always the case.

The Blue should be a self colour and there should be no lacing or barring, as in Andalusian Blue. Some strains appear to follow the imperfect colour syndrome producing Blacks, Whites, Black and White Splashes, and Blues.

There are some strains which breed true, although the extent of such strains has not been confirmed. The problem may be that the Blues, like Partridge, are not yet stabilized.

Sunshine and exposure to inclement weather will affect the shade of Blue obtained. Sometimes it becomes patchy and too light. Older birds may suffer from this defect and yet these will be more mature and better show birds than the young stock which is a fresher, evener colour.

5

MANAGEMENT

Courtesy: G Clayton

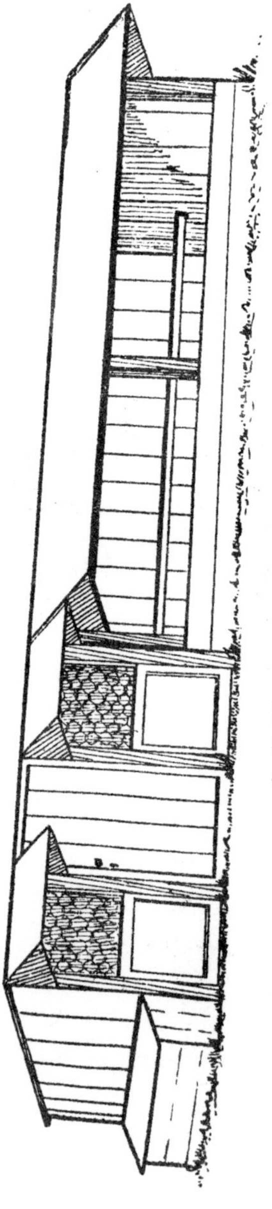

A Self-contained House

This may be used in all seasons

SPECIFIC SILKIE PROBLEMS

The Silkie has features which require special attention and therefore the intention is to deal with its management along the lines of covering these requirements. The main headings are:

1. **Housing & Essential Equipment**
2. **Feeding**
3. **Breeding**
4. **Health**

The first two are covered in this chapter and the others in Chapter 6.

Housing

The Silkie is quite hardy and therefore needs no heat or special requirements in that respect. The house should be established with the following rules being followed:

1. Size
Adequate for number of birds. A shed 1.5 metres x 2 metres should be adequate for about six hens, and up to 12 if there is a large run. However, do not over crowd. A smaller shed would house on a pro-rata basis.

2. Srong and Vermin Proof
Besides strong enough to keep out the elements it should also be rat and fox proof. Usually these can be kept at bay, but should

be safeguarded against and not encouraged by food being left around. If a wooden floor is to be used this should be placed on sleepers or flagstones; otherwise, in a very few years, the reaction of the birds - droppings, water and other elements, plus the earth underneath, will cause rotting and rats or mice will get through. Concrete floors are a sound idea, although cold. The answer is to cover them with shavings or soil to give scratching space.

3. Ventilation Adequate without Draughts

Poultry cannot stand cold, damp conditions or excessive draughts so care should be taken to put vents which do not blow directly on to the birds.

4. Pleasing Design

The experts usually specify that the design should have aesthetic appeal (pleasing to the eye) and be practical. These two requirements apply with great force with a garden poultry set-up. Nothing looks worse than a home made shed with corrugated iron sheets and badly planned layout stuck in the middle of a garden.

5. Screens and/or a Special Section

The fencing used can keep the birds within their domain and also landscape the area so that the poultry unit enhances the general

Garden Poultry Sheds

Ideal for Silkies because in inclement weather the birds can be confined. Top shed has extra run.

appeal of the garden. Trellis work, interwoven fencing, link fencing or other dividing material, interspersed with bushes, would be ideal. There could be a 'no-man's area' to separate the poultry unit completely and this should be located at the bottom of the garden.

Special Needs of Silkies

The Silkie is very hardy and will stand extreme temperatures, but its vulnerability are its legs and featherings on legs and toes. Accordingly, protection should be given so that the feathers do not become dirty or broken. Mud and dirt attaching may do considerable harm and, in addition, will encourage the scaley leg mite to attack the legs. The answer is to ensure that the floor of the shed and run are kept clean and dry. In the run wire netting can be used to give adequate fresh air, but the roof should be covered, together with sufficient overhang to prevent rain blowing in. There are many excellent products on the market to cover a wooden structure; if possible avoid corrugated iron sheets because they rust and do not provide adequate insulation.

If space and the budget allow a large shed with compartments, each with a wire netting front, and perspex or glazed shutters for winter, this is an excellent arrangement. A central inside corridor provides access with sufficient space for feeding under cover. Even a penning room for show training can be included.

Possible Types Of Building

There are very many types of poultry house which could fit into the garden layout. If an orchard or largish area of grass is available some kind of movable ark can be used, thus giving the birds the benefit of fresh grass which makes excellent food.

If space is limited the house will have to be relatively small with a run combined with provision for this to be covered in winter or it will become water logged. Some types of accommodation combine sleeping quarters and a run for scratching, all under cover. In that case a regular supply of greenstuff will have to be supplied to the birds.

The possible types are as follows:

1. **Small House with outside** run which is simply an area of garden fenced off and the birds are closed in each evening (a vital precaution against predators).
2. **House with internal run.** There are many clever designs available.
3. **Sussex Ark.** A popular shed for those who have a patch of grass.
4. **Fold Unit** which is a pen with a covered area at the end for sleeping and laying.
5. **Large Shed with Compartments.** This is mentioned in preceding section.

NOTE: ***There are many other variations which can be purchased or made. Detailed specifications of houses are to be found in Poultry Houses & Appliances – a DIY Guide .***

Various Types of Shed

The Choice

Once the possibilities are examined the appropriate shed and run can be made ready. Do not overcrowd, especially if there is limited scratching space. There should be adequate perch space and room for the food and water hoppers. If pellets are to be fed they should be kept under cover.

As a general rule each bird should have 2 sq. ft. of space in a shed with 3 sq. ft in an outside run. A typical shed (1m x 1m) plus a run of 2m x 1m should be adequate for six to eight layers, but ensure there is adequate perching space. When there is no outside run each bird should have 5 sq.ft of space. The height of the shed should be 6ft for each bird will require 25 cubic ft of breathing space.

A shed should be of the type that can be erected in sections, each bolting together. This facilitates removal and repair. Remember that birds can make the ground sour and therefore a move every two or more years will be advisable. Each year the outside run should be turned over and limed. If leaves and grass clippings are added regularly the run will be kept fresher and the additions are a source of food.

A soil which **drains well** is to be recommended, or the birds will very quickly get bogged down with mud. If birds are to be let out all winter in an open run try to have at least basic draining channels and near the pop-hole exit, it is helpful to place a thick layer of shingle

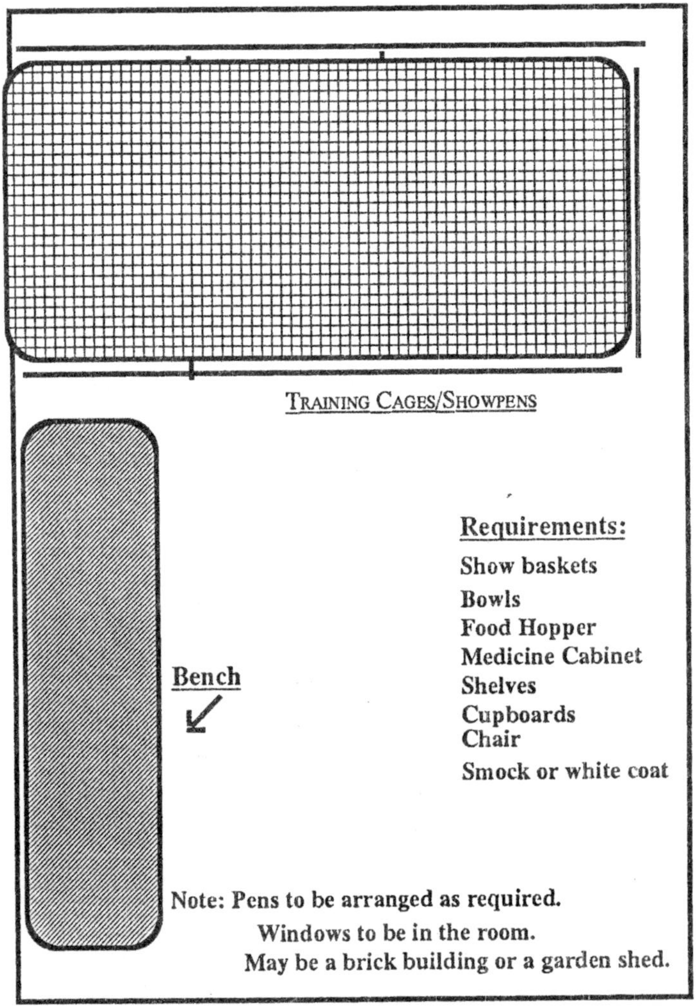

Plan for Show- or Penning Room

The type of metal cages used at shows may be used for training the birds; alternatively, wooden boxes with wire fronts may be employed (cage bird fronts can be utilized). Try to limit the period when the birds are kept enclosed with limited exercise. The purpose is to get the birds clean, fit and absolutely tame (until a bird will eat out of your hand). Then when being shown the exhibit will stand and be judged to the best advantage.

or pebbles so that mud is not carried in - this will soil the floor and dirty the eggs. A well known zoological garden keeps its Silkies on a concrete floor and this is washed down regularly, but this seems cold and unnatural, with no access to the earth.

Storage Bin

Food should be stored in a steel hopper. An old, disconnected chest freezer makes an excellent storage unit and this should be placed in a suitable shed. A dustbin with lid makes an excellent storage unit.

Try to avoid leaving food around or rats will be attracted. At the first sign of these creatures (evidenced by sight or droppings) put rat poison in a container with a hole for the rats to go in and eat the poison. It must not be put in any place where birds or pets can gain access to it.

SHOW REQUIREMENTS

If the birds are kept for exhibition, a Penning Room for training the birds will be essential. ***A layout is given on the preceding page.***

ESSENTIAL EQUIPMENT

Once the shed and run are ready the birds should be obtained and placed in the shed. For about a week feed and water them indoors and then, once they are settled they can be let out into the run. For the first few days of their release they may have to be driven in to the shed until they become used to going in.

The vital equipment for keping the birds will be as follows:

1. **Food hopper**
2. **Water container or fount**
3. **Nest boxes**
4. **Perches**
5. **Droppings Board or Tray**
6. **Grit hopper**
7. **Greens rack**
8. **Dust Bath**

Food Hopper

The best way of feeding birds is to have ad lib feeding with layers' pellets or mash in a hopper which is self-adjusting. We prefer the pellets because they are more palatable. Generally birds do not over-feed, but if they do the hopper can be opened in the mornings only and then closed at lunch time. The evening meal may be mixed corn scattered in the litter so the birds will scratch.

There are various sizes available so counting on 4 oz per day, per bird, place enough in the hopper to cover a week. Do not over-fill or the food will go mouldy.

Food hopper

Water fountain

Grit hopper

Food & Water Utensils

Water Containers

These can be bowls or other containers, but the most suitable is the water fount which regulates the flow of water and prevents dirt getting in the main container. A two-gallon (10litre) fount is suitable and this should be checked daily. Immediately water runs out wash the fount and replenish. Without water, even for a day the birds will suffer and may resort to eating their eggs, a habit which must be avoided.

Nest Boxes

Adequate nest box provision is essential. A breeding trio should have two nest boxes. If they are to be used as hatching boxes a hinged door should be provided so that when the Silkie decides to sit she is not disturbed by the other hen. If space permits a separate hatching shed is desirable so the hen get be let off for 15 minutes each day in complete privacy and then placed back; generally she will return herself without any difficulty.

Nest boxes are a source of vermin so they should be cleaned out regularly and new nesting material placed in at a level to make a nest. Straw or hay will be suitable. The nooks and cracks should be sprayed or dusted with insecticide, preferably when the birds are outside. Never let birds inhale the liquid insecticide or they may die.

Perches & Droppings Board

Poultry like to fly up as high as possible and roost and for that reason there should be adequate perches for them all to find a place. About 1 foot per bird is the minimum requirement.

The perches are best over a droppings board which is a shelf designed to catch the droppings; it should be covered in shavings and the droppings raked off each week. The ideal width is 2 inches (5cm) and the top should be rounded off. For ease of cleaning the perches should be placed in slots, U-shaped. When heavy breeds are kept make sure the perches are not too high or the birds may injure themselves.

If natural branches are available cut a suitable length of the appropriate thickness and suspend above the droppings board.

The droppings board keeps the faeces quite separate from the floor litter and thus gives more space; as a result the arrangement is more hygienic. If the droppings board is also removable this may be taken outside and washed down. It should be dusted with dustbin powder or washed with a mixture of water and Jeyes Fluid.

Grit Hoppers

Grit is essential in two forms:

1. Soluble grit such as oyster shell or limestone which provides calcium for the well-being of the bird and a coating to form the egg shells.

Perches are 2 ft apart and 1 ft above the droppings board.

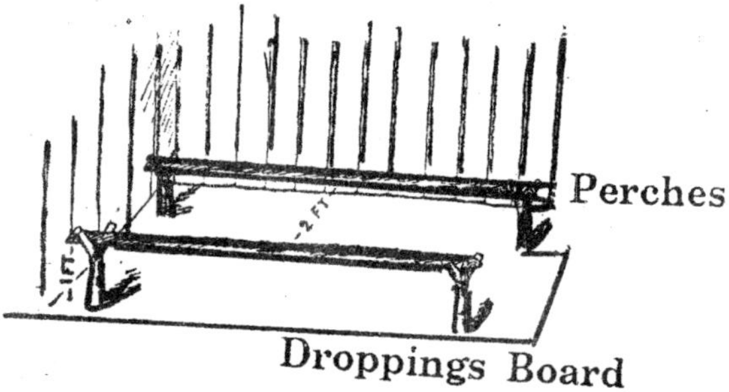

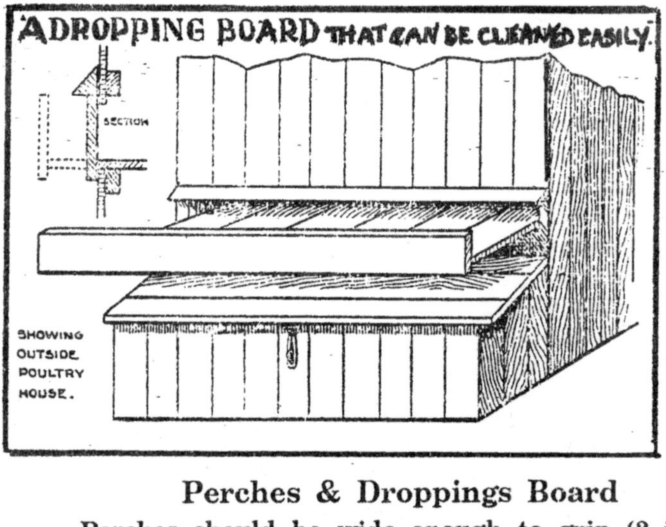

Perches & Droppings Board

Perches should be wide enough to grip (2 inch for large fowl) and slightly rounded. The top fittings are for supporting the perches.

2. Small pieces of flint grit to grind up the food when it reaches the Gizzard, which is a kind of organ which the bird uses to make the food digestible.

The hopper should be kept under cover so that the rain will not sour the soluble grit. The insoluble grit - the small flints - will be scattered in the run so the birds can pick it up when required.

When layers' pellets are purchased from the local mill they will contain limestone or similar grit so some is being provided. However, because corn may also be fed, it is wise to have other grit available at all times.

Greens & Greens Rack

A regular supply of greens is vital and therefore any birds which are kept in a closed run should be fed greens at least every other day. A rack may be provided or a cabbage may be hung from a post and the birds peck this as required. However, do not hang too high so the birds have to jump. Hens are not gymnasts and may injure themselves if the greens are too high.

Sprouting greens may also be used to supply the vital greenstuff. At one time there were special, metal cabinets available on which trays could be used to sprout corn. Within a few days, in a warm temperature, the corn sprouts and grows. It has to be watered each day and a nutri-

How to make a dust bath.

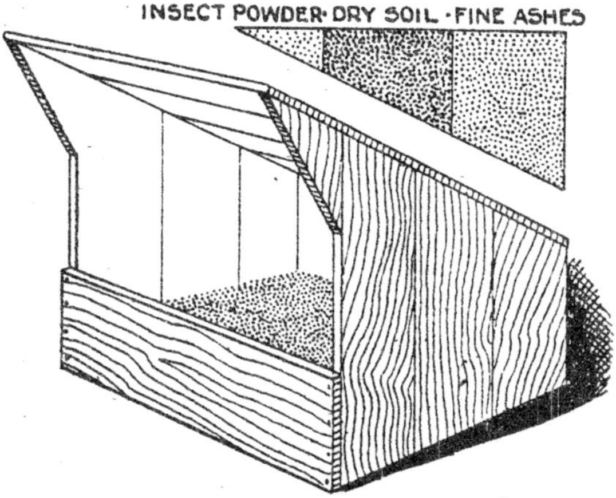

The Dust Bath

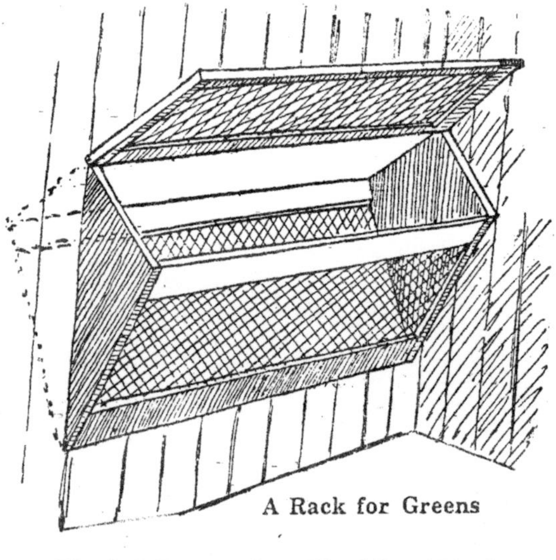

A Rack for Greens

Useful Items for Healthy Birds

ent solution can be added to encourage growth. A tray is then put into the pen and the birds will eat every morsel.

Even simply soaking grain in a bucket of water for 24 hours and then draining and leaving the grain to sprout can provide an ideal food which the birds eat very readily. It will help the growth if a sacking is placed over a tray of sprouting corn. In two to three days the corn will be ready to consume.

Mixed corn, including maize, is recommended because this gives a better vitamin coverage (Vitamins B and C) than wheat alone.

Dust Bath

Unfortunately, the warmth of the body of the fowl does attract scavengers such as mite. If these are to be controlled then the hens should be supplied with an area in which dry, fine soil is available in which the hen can dust herself. If space is limited then a box about 2 foot square can be supplied in which is placed fine soil, sand and fine ash. If really dry a hen will throw this up onto her back and roll in the dust. This has a cleansing effect and clears the mite.

An additional precaution is to pick up a bird and check around the vent; sulphur ointment very quickly clears away any mite.

Regular dusting with a special poultry delousing powder is also advised.

FEEDING

A silkie is a soft feathered bird, which means that it should have a profusion of feathers. Growing and maintaining the plumage at the desired standard requires a fairly high protein diet with vitamins included. Accordingly, it would be unwise to feed on wheat alone. A much better idea is to have *ad lib* feeding of layers' pellets, with extra greenstuffs, and the occasional feed of mixed corn. The latter would also be fed as the main diet when a hen is broody because this does not upset the bowels; avoidance of a hen soiling the nest is essential for eggs may be spoilt.

Do not feed on wheat alone because this is inadequate in vitamins and would not produce the profusion of silk required. Variety is essential.

Feeding

There are two schools of thought on feeding summed up as follows:

1. Ad Lib Feeding

The birds are supplied with a hopper which allows the mash (a powder) or pellets to run down into the base from which the birds feed. Greens can also be fed or the birds run on to grass, but for the main feed there is total reliance on the pellets or mash. Pellets are more palatable and cleaner, not clogging up the water from the birds drinking.

2. Hand Feeding Two or Three Times Daily

Pellets may be fed in the morning followed by scraps and greens at mid-day and then mixed corn in the early evening.

It is important not to overfeed for any surplus will encorage rats or mice and hens which are fed too much are in any case slothful and, once too fat, will lay very few eggs.

For the fancier this is the best approach, but when it is not possible (due to business commitments) a compromise will be necessary, such as having pellets in a hopper and once daily throwing a handful of corn per bird early in the morning or in the evening.

When exhibiting birds it should be appreciated that the hard feathered varieties such as Old English Game must have corn to attain the tight glossy coat of feathers; fed on pellets alone the birds become too feathery.

Watering

A regular check on the water will be essential. If time permits a daily change of the water will be necessary, especially when the weather is very hot.

In winter when the water is frozen it will be essential to empty the water fount in the evening and then, using a watering can, replenish with sufficient water for the day. If there is snow the birds outdoors will peck it to obtain the 'drink', but this should not be over done because fresh water is better. Sometimes, in very cold weather, the birds seem to appreciate warm water. If the founts are really frozen, hot water can be poured in the base so the water is freed.

If there is snow the birds should be kept indoors in a scratching shed. The dedicated fancier will ensure that suitable accommodation is available for this purpose.

There is no standard amount of water per bird per day, but to be on the safe side allow 1 pint or 1 litre per adult bird and make sure they always have plenty. If dry mash is fed more water will be taken.

Wet Mash

The feeding of a wet, crumbly mash can stimulate the appetite of birds and this was a common practice, but nowadays the ad lib feeding which is labour saving tends to be used. If a person has the time mixing a wet mash can be quite beneficial to the birds. Bread, scraps of cheese and meat, and other tit bits can be added, all waste from the household. However, it must be realized that the ration must be balanced and contain enough protein to sustain the body and egg laying.

Silhouettes

Trio of Black Silkies

Note the difference in size between cock & hens. Courtesy: T Penny

In protein terms we must think of the following:

1. Chicks - about 20 per cent.
2. Growers - 14 per cent.
3. Layers - 16 to 18 per cent.

Fish and meat meal is more effective than protein from peas or soya. Feeding more protein than stated is wasteful and if fed to Growers may bring the birds into lay **too quickly** which will reduce the size of the egg.

The premixed food from the feed mill or poultry food merchant shows the level of protein, this being required by law. Corn has a protein level of around 10 per cent so it will be apparent that a complete diet of corn is not advisable, but it does delay maturity. It must be supplemented with greens and tit bits.

Purchasing the Feedstuff

If six or eight layers are to be kept they will eat about 50 kilos in 6 weeks and this should be purchased in bulk. Buying small quantities is not economic. Once an initial stock is acquired, In the interests of economy, a bag of mixed corn may be purchased every 3 weeks and the layers pellets 3 weeks later, thus budgeting for the cost. Make sure there is maize in the corn, because wheat alone is inadequate. Do not over stock because the pellets may not keep, especially in damp weather.

Feeding the Best

Hens can be fed all the household scraps and will love them, but for maximum production remember that these should be limited to not more than 20 per cent of the total intake. Thus bread, soaked in water or spare milk, cheese, rice, meat scraps and other tit-bits will all be consumed and the birds will become quite tame when being fed. However, do not feed mouldy bread or contaminated food; this may cause problems and taint the eggs.

Fowl have a **feeding cycle** which is along the following lines:

1. The Sitting Hen

She eats little food when she is sitting and provided she has corn and water each day is healthier from sitting.

2. Chicken-hood

The small chick eats nothing the first two days and then gradually increases the amount, but in the first two months costs very little to develop. However, the food must be sound and contain sufficient protein. Do not hope to rear on scraps because this will not pay. Also chick crumbs from the local mill will contain a medicine which protects against coccidiosis, a condition caused by a germ picked up by the chicks running outside. Generally, though chicks are healthy and require little attention other than than good food, adequate fresh water and warmth.

Grit

There should be a regular check made on grit available, both flint and limestone. **Without grit the birds cannot function properly.**

At the first sign of soft shelled eggs or shells which are uneven **take positive action.** The problem may be a shortage of grit or the hens are too fat or they are not taking the grit. In the pellets purchased from the mill grit is included, but some hens may need more, but will not pick it up. In these circumstances sprinkle fine grit in a wet mash until the eggs are well shelled. If the problem persists try to spot the hen and then feed her grit by hand - a little fed into her beak will suffice.

Soft shelled eggs are a menace and must be eliminated. The eggs are of no value and can lead to egg eating, which is difficult to stamp out.

Storage

As noted earlier, there should be adequate provision for storage. A metal hopper or even a dustbin will suffice to keep the food fresh and away from mice.

A Store Room combined with a **Penning Room** to train birds or carry out other functions connected with the birds will be found to be a great help.

Culling

Health and condition can be seen from handling. The body should be firm, but not fat. If the vent is inspected it will be found to be large and moist. If two fingers can be placed between the pelvic bones the hen is usually laying. However, if the space is very narrow it means no eggs are being laid and if this persists the hen will have to be culled; ie, disposed of.

At this inspection check for mite at the vent and dust with insect powder. If the feathers at the vent become soiled remove them and rub the area with a light coating of sulphur ointment which kills any mite. If there are clusters of tiny eggs pull these off before applying the ointment. Sometimes the feathers on the back become very worn and broken due to an over enthusiastic cockerel. In this case he may need more hens or he should have two pens of birds, letting him to each on alternate days. Again, treat the area with sulphur ointment.

If a hen becomes damaged through any cause she should have the area bathed and a medical disinfectant added, such as Dettol. In the case of a bad injury or she is being pecked, separate the hen from the main pen until she recovers. A small coop and run is adequate for this purpose.

If, on inspection a hen is permanently on the nest, then, if not required as a broody hen, move her to a special pen with a wire or slatted floor which will break her of the habit. Alternatively. place her in a new run with a different cock and this

Periodic Tasks

The runs should be kept in good repair and when foxes are in the area make sure the birds are locked up each evening. Wire should be sunk into the ground so that no tunnelling is possible. This should be checked at regular intervals.

Door hinges and locks should be oiled and the sheds should be painted or creosoted about once a year. If mite is to be discouraged the perches and droppings boards should be removed and all crevices should be painted with paraffin or creosote. This can be done before the breeding pens are made up at the beginning of the year.

Splendid Champion White Hen

Courtesy: W Heaton

Lighting

If possible a small wattage bulb should be used to give the birds light (to bring up to a total of 12 hours per day). Usually production is increased by around 20 per cent by giving the extra light so the exercise is very worthwhile.

A time switch can be used or the light can be switched on in the evening and then turned off before going to bed.

Weekly Routine

There are many tasks which will have to be done at the weekend or on a freeday. These include the following:

1. Renew Litter & Nest Box Material

The floor of the poultry house should be strewn with litter to a reasonable depth. This should be checked each week and any soiled material removed and replaced with new.

The best material is dry peat moss or wood-shavings. The latter is cheaper, but does not mix so well with the droppings to make manure. Straw and hay may also be used, but they do not absorb the moisture to anywhere near the same extent. These can be used for nest boxes. The natural material is soil so if this is added on a regular basis the birds will scratch in it; they also eat it.

2. Clean Droppings' Board

If a droppings board is placed beneath the perches this should be cleaned by removing the faeces and sprinkling sawdust, peat moss, sand or dry earth. Remember this is excellent manure for the garden, but it is also quite strong and is best allowed to dry **under cover** before using.

A scraper may be found quite useful for ensuring all the droppings are removed easily.

3. Pick up Birds and Handle Them.

This might be best done in the evening when the birds have gone to roost. If they have to be caught up use a large net and if they escape do not chase them too much because this will induce alarm and may cause them to stop laying; open a pen door and gently drive in. A Silkie should feel plump and quite furry; lack of feathers is a serious fault.

4. Gradual Growing Period for Laying Stock

Pullets should be brought on gradually and must not be over fed. It has been found that if fed too much the pullets will start to lay at 5/6 months of age and the eggs will be too small. Therefore avoid over feeding and limit the protein; Growers pellets are suitable.

Partridge Silkie Hen
Best Coloured Silkie Hen at London Championship Show
(Courtesy: Jane Holroyd)

Note: This well presented bird is an excellent specimen, its only apparent fault being the flatness of the cushion, and slight hardness of the tail feathers.

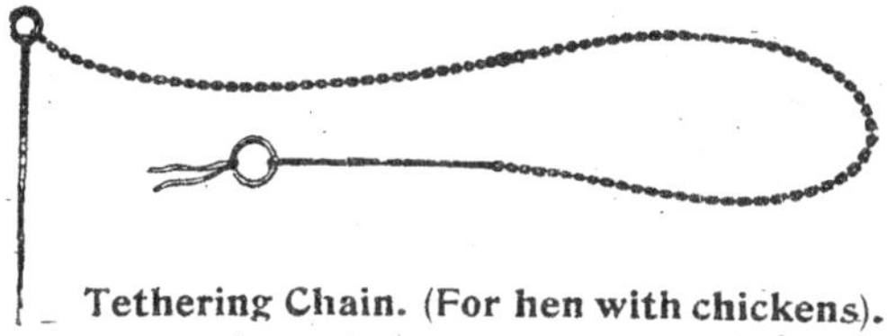

Tethering Chain. (For hen with chickens).

The skewer is pushed into the ground and the tape (on rubber ring) is wound several times round the hen's leg to secure her. The rod prevents the chain winding round the hen's leg. A length of twine with a piece of wire for a rod answers as well, and costs but a few pence. A swivel on the chain or string is necessary.

Tethered Broodies

Can be used where there is privacy and safety from cats, foxes and other predators,

6

FURTHER MANAGEMENT

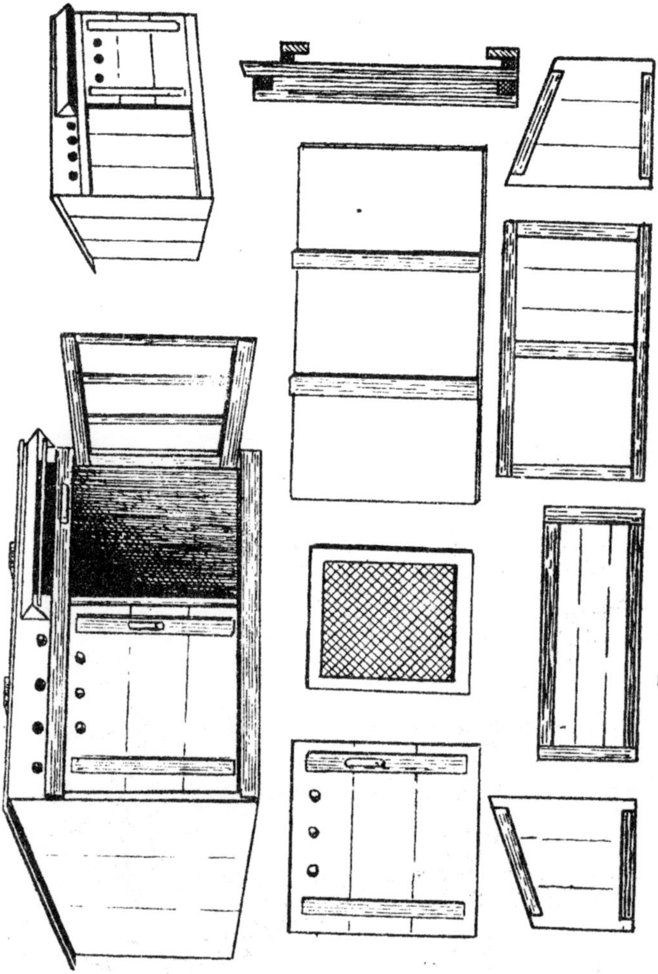

THE DOUBLE COMPARTMENT COOP AND ITS FITTINGS

Top. The coop shown with the front open and, on the right, closed for the night

Centre. The wooden shutter; the wire-netting front to fit over the bars in warm weather so that the wooden shutter may not be used; the roof, the centre section of which is flanged to form the door, the other parts being screwed down; and a section showing the slides for the shutter

Bottom. All the framework which is required for the ends, the back, and the front. The coop should be fitted together with stout screws

Double Broody/Rearing Box

Shows sections and how it may be made.

MATTERS TO BE CONSIDERED

In the preceding chapter basic management is considered, with special reference to housing, essential equipment and feeding. The process is taken further and the two other main aspects, so far as they affect the Silkie, are now examined.

The main assumption made is that the Silkie is to be kept primarily as a hobby and, as in the previous chapter, the usual management will be on a small scale in a garden, and trellis work, interwoven fencing, link fencing or other dividing partitions will be used. Breeding will usually be by means of the birds themselves hatching the eggs. However, if a number of breeding pens are kept the fancier may decide to use an incubator, thus enabling more chicks to be hatched. It will be appreciated that with the best will in the world a broody hen will sit on about 12 eggs and she will be occupied with the eggs and the resultant chicks for 8 weeks, possibly longer, which will eliminate egg production for that period. In such circumstances, the hen may be moved around to different pens and, if necessary, to a Coop for breaking broodiness; this consists of a box with a wire netting or slatted floor and an open front which will discourage the broodiness. During that time she should be fed layers' pellets and given water and generally she will start laying again fairly quickly. If a second clutch of eggs can be obtained before allowing her to sit this will increase productivity by 100 per cent.

Unlike other breeds of poultry the Silkie lays a clutch of eggs and immediately becomes broody. This clutch may be as many as 18 eggs, but may be fewer or more, depending on the strain. It is very unlikely to be more than 20 eggs before she shows signs of sitting.

The Breeding Pen

A male bird is not essential when no breeding is intended from the birds being kept. However, if the poultry keeper wishes to obtain replacements, or if showing, to breed winners, then it will be essential to have a breeding pen.

The cockerel or cock (over a year old) selected should be active and full of vigour. Generally, in the breeding season, he should also be attentive to his female companions and scratch and find food for them. Two year old hens are best.

Remember the male is very important; he is half the pen and if his background is doubtful many of his faults will be transferred to his offspring. This is why some breeders of show birds advocate 'line breeding' or even 'inbreeding' so that related stock are bred together, thus locking in all the desirable characteristics. Thus a son **may be matched to the mother.**

Both sides of the breeding pen should be of the correct size; with the standard bred birds (British) they should be as large as possible, but not coarse or lacking the essential cobby shape. There should be very full plumage (lots of silk) and no hard feathers.

Try to avoid males with sickle feathers which are prominent and are fully webbed. This is a serious fault so there is little point in breeding from such birds. Regretfully, many coloured birds (other than white) suffer from this problem, a sign that there is foreign blood in the make-up of the variety.

Faults, listed in the standards, should be noted and if a bird possesses one or more of these to a marked degree avoid breeding from the birds in question. Typical of these are **vulture hocks, featherless legs**, white skin, **single comb**, legs not of a lead grey colour, red faces, **lacking the fifth toe**, green colour beak, green on soles of feet, **green legs**, faulty crest and hardness of feathers.* The birds should lively and active, constantly on the move, but not as excitable as some of the light breeds.

Mating the Cockerels

Usually up to about 6 hens to one cockerel is permissible. Chicks should be hatched in the period February to the end of April.

For **show birds** the breeding pen may include **a cockerel and two hens (referred to as a breeding trio) and in extreme cases it will be one** to one, possibly the son to the mother or the sister to the brother. Early in the year use very few hens.

*** Those in bold should call for disqualification, but others in sufficient numbers may also; eg, wrong colour flesh, incorrect colour.**

Collecting & Storing Eggs

Eggs should be collected at least once per day and if possible they should be clean when taken from the nest box. If not clean they will have to be washed on the same day, before any germs have accumulated on the egg. If for eating they would be stored in a refrigerator. For hatching they should be stored in a cool temperature in a tray or box containing shavings.

If for hatching the eggs should be marked with the date and the pen number. When washed a mild disinfectant should be used, thus giving them protection.

Each day they should be turned and not later than 7 days after laying batches should be placed in an incubator and hatched. Alternatively, a broody hen may be used. Remember that the incubator is made to run at least half full so a reasonable number of eggs should be obtained before setting.

Hatching the Eggs

Hen eggs take 21 days to hatch and bantams a day or two shorter. They should be set in the incubator* in a room which is well ventilated and not subjected to wide variations in temperature caused by the sun shining through the win-

---

* For full details of hatching & rearing procedures readers are referred to *Natural Incubation & Rearing* and *Artificial Incubation & Rearing*, Dr J Batty

dow or other reason. There has to be the correct combination of air, heat and humidity for the eggs to hatch. Any critical variation will not perrmit eggs to be hatched.

The instructions of the manufacturer should be followed, but usually a temperature of 99.6 degrees F (37.5^{0} C.) is the aim. In the last week the temperature may be allowed to drop slightly and humidity increased. In effect, the actual egg is at about 100 degrees so the thermometer above the egg should be higher.

Eggs are turned twice per day and at 7 and 14 days the eggs are tested by holding up to a bright light (a torch will suffice); this is known as 'candling'. Any eggs which are quite clear are removed because these are infertile (indicating a fault in the breeding pen) and those which show a growing embryo are placed back in the incubator.

At 14 days, any eggs which show indistinct, blurred outlines, should be regarded as dead embryos, but they should not be rejected unless the operator is experienced so a double check will be essential.

The Broody Hen

When a broody hen is to be employed a mature hen should be used (pullets do not always have the stamina). When she is seen staying on the nest at night and you require her to be broody, place some pot eggs under her for a few days and place your hand under her when, if broody, she will peck and clutch the eggs with her wings and body. If she flies off in fright then it is likely that she is not yet ready. Fortunately, the Silkie does not generally show any wayward tendencies because broodiness is second nature to her.

Once broodiness is established then transfer her to a broody coop on a pre-made nest of earth lined with straw and place the eggs beneath her. At times, getting a broody to settle is difficult because if she is moved she may desert her new nest. One plan is to have a nest box with lid, which can be moved to a private shed (usually at night time with her on the eggs), and in this way she does not realize that she has been moved. This privacy is essential because, if- left in the communal house, other hens will disturb the broody, laying eggs in her nest, so different aged eggs are mixed together.

The Silkie is the best broody, but she is on the small side and her lower breast feathers may get tangled and a small chick get caught and asphyxiated. A cross bred (Silkie X Wyandotte is excellent) is larger and avoids the difficulty. The breast feathers should be plucked to avoid tangling.

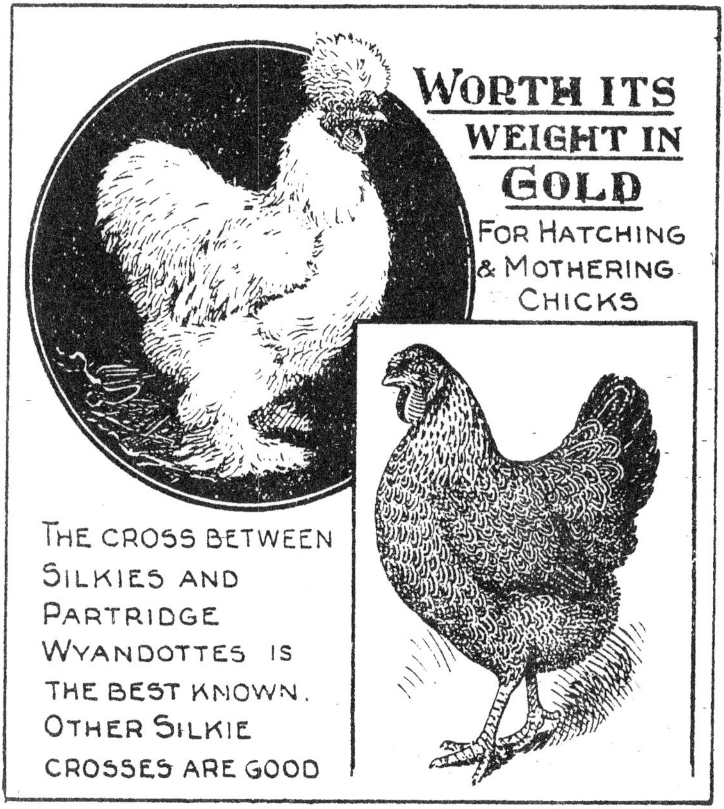

A Broody Hen

May be Silkie or Silkie X Wyandotte

The broody should be removed once each day for toilet, food and water. If this is not done she may foul the eggs and unless fed daily on mixed corn she cannot sustain the 21 days confinement.

If she is dusted with insect powder **before** being put to sit there should be no problem with mite, but keep an eye on the posibility by inspecting the vent at 10 days.

Leaving the Chicks

Once the chicks start to hatch do not disturb them until dry. Only then should they be placed in suitable quarters. They require no food for 24 hours although the broody hen should be given corn and water when the chicks are moved.

REARING

Rearing chicks requires great care and attention, especially during the first week. Unlike many other young birds chicks are quite independent and can peck for food from about 2 days' old. They need warmth (90 degrees F reduced by 5 degrees each week until around 65 is reached), chick crumbs and water in a small, chick fount. Broken corn and chick weed can be given as well.

Silkie chicks tend to be delicate so great care should be exercised for the first two weeks. Rearing in wooden cages with wire fronts (cage bird wire fronts are excellent) produces excellent results. This may be by using an infra red lamp or a Silkie broody.

Rearing By Broody

The broody hen will watch over the chicks and coax them to take food and water. Generally the naturally hatched chicks are very healthy and strong. She may be kept in a coop with a run attaced which can be moved on to fresh ground every 10 days or so. The grass should be kept reasonably short.

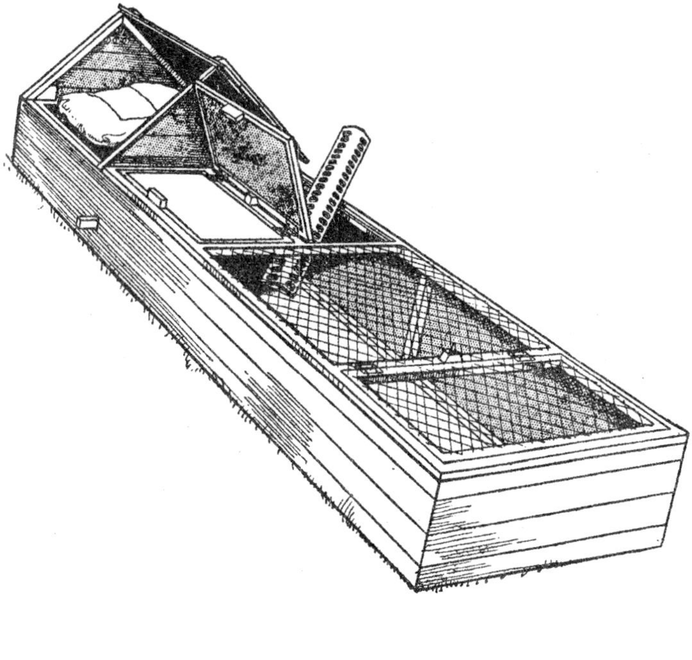

A Broody Coop for Chicks

Rearing cages are an alternative and avoid disease better

Growing Period

Chicks can be transferred on to Growers' mash or granules from about 3 weeks of age. Introduce this gradually, but keep feeding the broken corn.

At 7 or 8 weeks of age the mother can be put back to lay. Often the hen will show signs that she is ready to return to the breeding pen by pecking the chicks - Nature's way of ensuring that the breeding process is continued. Make sure the hen is not bullied when she is returned to her pen.

Those reared artificially will be ready to go without heat and should be placed in a small shed with run so they can grow in the sunshine. A lining of hay may be helpful to keep the chicks warm until fully hardened. In very cold, frosty weather keep them indoors. A rearing pen in a large shed will be ideal.

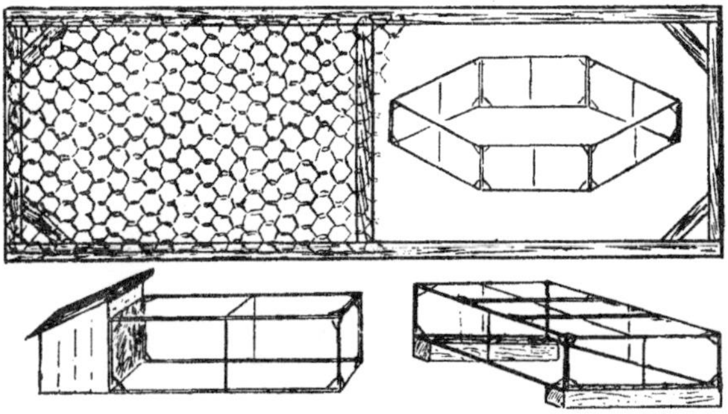

A USEFUL FENCE FOR CHICKENS

Fence units are very useful during the rearing season. Convenient sizes are 6 ft. and 2 ft. long and wide enough to take the width of 2 ft. netting. Mine are made of 1 in. square deal, braced at the corners, and the 6 ft. length braced through the middle also, the frame being creosoted before the netting (1 in. mesh) is fixed. Three 6 ft. lengths with a 2 ft. end make an excellent covered run for a hen cooped with her chickens, while half a dozen of the long frames used end to end form a suitable enclosure for growers.

HEALTH

Silkies are long lived and are very hardy. Accordingly, generally they are very little trouble.

Because of their woolly coats it is sometimes suggested that they suffer from **mite** more than normal. This has not been the experience of the author, but it is admitted that cleanliness must be observed. Moreover, dusting powder should be used and the occasional clearing of feathers from around the vent, followed by an application of a small amount of sulphur ointment will suffice to keep them free. When washing for shows a little disinfectant in the water will also keep away feather mite, but avoid allowing this to get in the eyes.

A more serious problem is **scaley leg** caused by a small parasite that gets under the scales and disfigure the bird. The scales become raised and uneven and leg feathering suffers. The birds are unfit for shows and they become carriers of the condition.

As a boy the author remembers his father using a mixture called 'Troopers Ointment' which was very effective, but is no longer available because of poison regulations. Soaking and washing the feet in hot water containing disinfectant and a liquid soap can be the start of the cure. After that the legs are rubbed thoroughly with sulphur ointment or this mixed with medicinal tar. It may be obtained from a local pharmacist.

Silkies appear to be susceptible to this condition, possibly because the feathering on the legs encourages the accumulation of dirt. The only way to avoid the problem is strict cleanliness of the legs and feet, washing regularly (say, once per month) with warm water containing washing-up liquid and a mild disinfectant.

Marek's Disease

A disease that has caused considerable problems to ***some strains*** of Silkie is a form of Fowl Paralysis known as ***Marek's Disease***. Early authorities believed that paralysis resulted from lack of greenstuff in the diet, but this is not a general conclusion, although possibly it would affect the disease. This is a condition which was quite commonplace in poultry, but is now no longer a serious problem and, therefore, it is surprising that the Silkie should be prone to its effects.

A vaccine is now available to combat this serious disease and has been quite effective. Since the malady is caused by a herpes virus the vaccine, administered at a day old, is a Herpes Vaccine.

There are various signs for spotting the disease. Lameness or a drooping wing are often the first indications of the problem. Tumours also develop and these bring about the deterioration.

Adult birds may recover, but often do not

so unless very valuable stock, should be killed and incinerated.

Why Silkies are affected is not apparent, but the disease is present in the cells lining the follicles of the feathers. Because of the fluffy nature of the breed is it prone to infection?. This is not clear and more research is needed to be positive.

What is certain is that the onset of the disease appears to be linked with nervous stress, and the danger period appears to be as they approach lay.

Coccidiosis

Brief mention must be made of the disease Coccidiosis which is a condition of the intestines, and which affects chicks in particular. It is transmitted by oocysts in the droppings of other birds so great attention must be paid to cleanliness and hygiene. This is why many breeders rear Silkie chicks inside in box cages and change the litter quite regularly.

Once fully feathered the chicks can be put outside, but on clean ground and preferably not in wet conditions because this seems to aggravate the condition. In the first 8 weeks special Chick crumbs, containing a Coccidiostat should be fed.

Note: For those who wish to consider in detail the various poultry diseases *Poultry Diseases Under Modern Management,* is ideal. Usually the avarage fancier gets no serious problems.

THE FRIZZLED AND SILKY FOWLS.

An Illustration from the Past

7

CULT OF THE SILKIE

Silkie Club Annual Report

1932.

The entries at the Club Show came up very well—two more than last year, and within four of the number demanded, our guarantee is thus lighter than it has been for some years.

The wet autumn has not been conducive to birds getting through the moult, and many good birds were very sappy. The Blacks have greatly improved, as have also the "Colours," there being 18 entries in the latter class.

A Points Cup is to be competed for at all shows, during 1933, where classes are guaranteed or recognized by the Club. Open to all paid up Members.

Mr. R. L. Fairley has given two medals for best White and best Black or Coloured Silkie, which will be competed for at Bradford Bantam Show in January. He has also guaranteed classes at Paisley. It is hoped these classes will be well supported.

Mr. H. W. Hall has offered a medal for best Novice at Club Show who has obtained eggs or stock direct from a Member.

Many expressions of regret were heard when it became known that Dr. Campbell, our genial secretary for 30 years, had resigned owing to health reasons.

The Club Show for 1933 is arranged for the Crystal Palace—Judge, Mr. C. A. House.

The Club has a membership of twenty-four and has a balance of £15 1s. 4d.

The following were elected at the Annual Meeting :

President : Mrs. A. M. Hall.

Vice-Presidents : Miss Chisholm and Mr. Draper.

Committee : Miss Lambert, Mr. Fairley and Mr. Fletcher Hearnshaw.

Hon. Secretary and Treasurer : Mrs. E. S. Fentiman, "Haldon," 186, Whitworth Road, Swindon, Wilts.

CLUB SHOW, CRYSTAL PALACE, 1932.

BREEDING PEN (7) : First and Cup, Miss Michie ; 2nd, C. F. Jones ; 3rd and Reserve, Mrs. Hall ; V.H.C., Fairley ; H.C., Draper : C., Watts.

WHITE SILKIE COCK (7) : First and Cup and V.H.C., Mrs. Hall ; 2nd and 3rd, Fairley ; H.C., Miss Lambert.

WHITE SILKIE HEN (7) : First and Cup and Champion Cup, Mrs. Hall ; 2nd, Mrs. Watts ; 3rd, Fairley ; Reserve, Mrs. Hall ; V.H.C. Mrs. Ashby ; H.C., Fairley ; C., Mrs. Hall.

WHITE SILKIE COCKEREL (9) : First and Cup and 3rd, Mrs. Hall ; 2nd and Reserve, Misses Davidson and Chisholm ; V.H.C., Miss Fleming ; H.C., Miss Michie ; C., Miss Lambert.

WHITE SILKIE PULLET (9) : First and Cup and 3rd, Misses Davidson and Chisholm ; 2nd, Miss Michie ; Reserve and H.C., Mrs. Hall ; V.H.C. and C., Fairley.

BLACK SILKIE, Male or Female (9) : First and Cup, 2nd and 3rd, and H.C., Draper ; V.H.C., Fairley ; C., Mrs. Hall.

A.O.C. (18) : First and Cup, and 2nd, Mrs. Hall ; 3rd, Reserve, Draper ; H.C., and C., Hassal ; V.H.C., Fairley.

An Announcement from the Silkie Club, 1932

Introduction

The Silkie has had a long and progressive life in the Fancy. From the very early days of poultry shows it has been in evidence and has grown in strength. At times the number of members of the Silkie Club has been small, but, from reports in the Fancy journals, there has always been enthusiasm, with fanciers such as Dr and Mrs Campbell, and other officials, who have worked hard for the cause.

This chapter deals with some of the significant aspects of the development.

Early Days

The early references confused the Silkie and other types of bird, which had Silkie characteristics. For that reason we cannot be certain when the breed in its modern form came into existence.

What is certain is that the early fanciers were also confused. There was so much uncertainty that they were not listed in the first British standard of 1865, and appeared in the American standards in 1874, both in the Bearded and Non-Bearded types, white in colour, *as bantams,* not large fowl. The American breed was obviously smaller than the British birds, yet from early times the British Silkie tended to be large fowl size (although small). Lewis Wright had asserted *

***Illustrated Book of Poultry*, which was written before 1900.**

that birds had been 4lb for cocks and 3lb for hens so this would make them larger than bantams, a position recently clarified by the Silkie Club, establishing these larger weights and also recognizing bantams as separate varieties.

There is also the question of birds which had Silkie-type plumage, and this occurred in many breeds. Even today there are examples to be seen in Japanese bantams (see illustrations).

Two old engravings indicate that the early Silkies were different from the modern birds. They were longer in the body and the 'silk' was not as profuse as the birds seen today.

The first engraving was produced in 1807 and shows the breed as having long feathered legs and a rather racey body. Although the plumage is silkie-type, it fits closer to the body. The shape is conventional, clearly showing the more upright carriage of the Jungle Fowl, as evidenced by the "Wild cock and hen", also illustrated.

The second engraving shows exhibits at the West Kent Poultry Exhibition in 1853. The surprising fact is that the 'Silkies', although a more cobby shape, did not have feathered legs or five toes. Moreover, the silk was certainly very poor by today's standards. The discerning reader will note that this date is well before the date given earlier of 1872 for the first listing at a show at Crystal Palace (p.17). Accordingly, it can be claimed that Silkies were in the Show Scene right from the very beginning of the exhibition era.

Japanese Bantams Silkie Type

An Early Engraving (1807)

Linn. Syst.

ORNITHOLOGY.

Div. 2d. Land Birds _Order GALLINÆ.

GENUS *PHASIANUS*.

MR. BAKER'S SPANGLED POLANDS. , 1853.

Silkies at the West Kent Show, 1853

Other sources confirm the existence of the Silkie. Lewis Wright (ibid) recorded that Silkies had been in breeders' hands from quite early times. The first reference he makes is to comments by the Rev. R S Woodgate of Kent who had been involved in the breed for 40 years prior to 1900.

He noted that a Captain Finch brought home a pair about 1860, probably from China. Another fancier, a Miss Hawker, imported birds from Japan in 1869.

One fancier made enquiries to discover whether the breed existed in Japan, but the reply back indicated that they could not be found. However, a report from another source confirmed their existence in Yokohama, but not in Kobe or Nagasaki. It appeared that they were located in some areas, but not others.

Lewis Wright also commented on the *quality* of the breed at the beginning of the 20th century, remarking that they were now shorter in the leg. He also went on to state:

> **We must not, however, forget that Silkies should have tails. I am afraid that many are sliding into a pure Cochin or Pekin bantam-shaped bird. This should not be. The tips of the tail feather, as silky as possible, should protrude beyond as dense a mass of feathers as possible.**

Modern fanciers should note this fact because some birds now being shown are very low in the tail!

Therefore almost a century ago Lewis Wright was giving guidance on a law of nature; namely, that in the constant search for more ***silk***, it must be appreciated, that growing a proper tail must become more difficult.

The Silkie Club of Britain

This is one of the oldest clubs in existence, founded in 1898, and continues to give support to the breed and the Fancy generally. The founder was Dr A Campbell, assisted by his wife Mrs Campbell*, a leading poultry breeder of the day.

The present Secretary is Mrs S Flude, Olde Barn Farm, Schooll Lane, Stapleton, Leics. LE9 8JR, and Honorary President Mrs D Bryer, active in Silkies for a very long period (34 years as Secretary).

The revising of the *Standard* is the responsibility of the Club, as well as arranging various activities and shows.

An early report from the Club is shown earlier in this chapter. Today the much increased membership is around 150, but it must be appreciated that this is only a guide to those dedicated fanciers who show Silkies. Many keep birds to act as broodies and these may never show, but continue to be interested in the breed.

*** I have not been able to verify with absolute certainty that the Mrs Campbell of Khaki Campbell fame and leading breeder of Sultans, Silkies and other breeds, was the same person as the wife of Dr Campbell, but the facts available support this view.**

Research on Silkies

In the past poultry researchers have been fascinated by the Silkie and its many unusual features. The black skin and flesh, the silky plumage, the unique comb, the five toes, and other characteristics, enable crosses to be made to determine whether a particular feature is Dominant or Recessive.

As will be appreciated, the first cross (usually referred to as **F1**), may not show what can be expected long term. It becomes necessary to make further crosses, breeding back to the original, before any valid conclusions can be reached. Creating new breeds or improving them is very much a long term project. This is shown very clearly by the work of such pioneer breeders as W F Entwisle who was responsible for producing many of the bantams available today. According to Sir Edward Brown*:

> **......to watch the work of the late Mr W F Entwisle, who was responsible for the evolution of more breeds of bantams than any other man, and their development. That gentleman's skill and patience were phenomenal. In some instances he spent ten years in perfecting a new breed. He was one of the greatest breeders of his day.**

The subsequent generations (F2, F3, and so on) will produce results which indicate the nature

---

***_Poultry Breeding & Production,_ London, 1929. The original standards formulated by Entwisle are given earlier and were taken from _Bantams,_ showing the work of Entwisle.**

of a characteristic.

Charles Darwin and Others

Many scientists have worked on heredity theories. Gregor Mendel* gave the foundation on which much of today's work is founded, but it was Charles Darwin++ who explained how the various factors came together and produced the results obtained.

The first reference to the character of Silkie plumage was by Charles Darwin (1868) who observed that Silkie X Silkie produced the same type (F1 generation). However, once crossed with normal plumed birds, the standard type of feather predominated; in other words, the Silkie plumage is recessive (F2 generation).

Others followed and confirmed the tendency. They included Davenport (1906), and Bateson and Punnet (1908). They were followed by Jull and others (1927 onwards), Hagedoorn (1953) and Stevens (1991) who classified silkiness as autosomal and recessive.

In the F2 generation it was found that around two-thirds would have normal feathered plumage and the balance would be Silkies. The result is brought about by the existence of an autosomal, recessive gene.

See *Mendelism*, R C Punnet, London, 1911. Mendel died in 1884, but his work was not discovered until 1900.
++ *Origin of the Species*, London, 1959.

PULLET
2nd Olympia 1924

PULLET
1st Olympia 1924

BELOW
HEN
Left:
Ist & Cup, Olympia 1925

HEN
Right
3rd Palace & Dairy 1924

Development of Silkies

It is interesting to compare these 1920s Winners with Modern Birds
These above give the impression of being smaller and not so rounded.

PULLET.
Reserve at both Palace Club Show and Olympia, 1924.

COCKEREL.
1st Olympia, 1924.

COCKEREL.
4th Olympia, 1924.

COCK.
3rd Olympia, 1923; 4th Palace Club Show, 1924; 4th Olympia, 1924.

A Team shown by Ian B Gow in 1924

The Work of Professor R C Punnett

Much original work on poultry genetics was carried out by Professor Punnett at the University of Cambridge. In connection with Silkies he confirmed the recessive nature of the plumage; he also experimented with different comb types and the results of crosses with Silkies

The presence of black pigment and its possible transfer to other breeds was another area of research. The results were quite varied dependent upon whether a Silkie cock *or* female was used on one side. With Leghorn crosses it was found to depend upon which variety was used for the cross. Not all accepted the dark pigment to the same degree.

A wing feather and a contour feather of an ordinary and a silky fowl. The peculiar ragged appearance of the silky feathers is due to the absence of the little hooks or barbules which hold the barbs together. The silky condition is recessive.

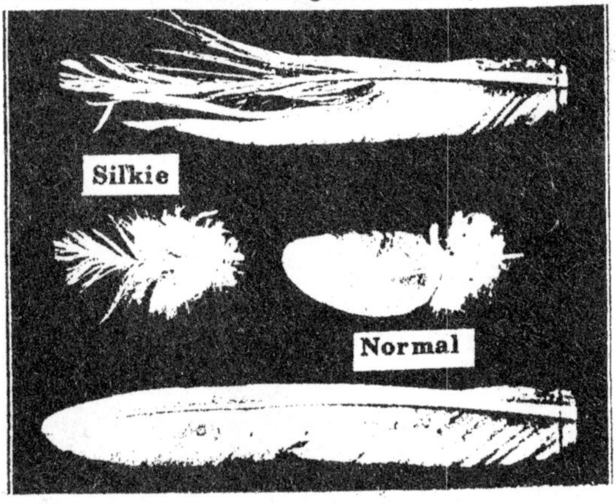

Feathers Compared (Punnett)

See *Mendelism*, London, 1911

Typical Rhode Island Reds
(male & female)
These are far removed from a Silkie in shape and character.

Research of A L Hagedoorn & G Sykes

The Hagedoorn and Sykes team were concerned with combining the theoretical and practical aspects of poultry keeping.*

One experiment involved a cross between a Naked-neck Rhode Island Red and a Silkie. The mating produced Black-red hens with naked necks. These were mated back to non-booted Silkies and the offspring were 50 per cent white and 50 per cent coloured.

In plumage the offspring were well mixed, some with normal feathers and others with silkie type. Two were up to show standard for Silkies.

The authors were of the opinion that a further cross back would have produced booted Silkies.

Note: This is an interesting case study, but readers need to know more facts before reaching positive conclusions. Obvious queries are:

1. Since a Rhode Island Red is not normally a Naked Neck type why was this used?

2. Silkies normally have feathered legs (booted) so why were clean legged birds used in the second cross? Was it that no feathered leg birds were available, which would be surprising if a pure bred Silkie was used initially.

***Poultry Breeding*, A L Hagedoorn & Geoffrey Sykes, London, 1953.**

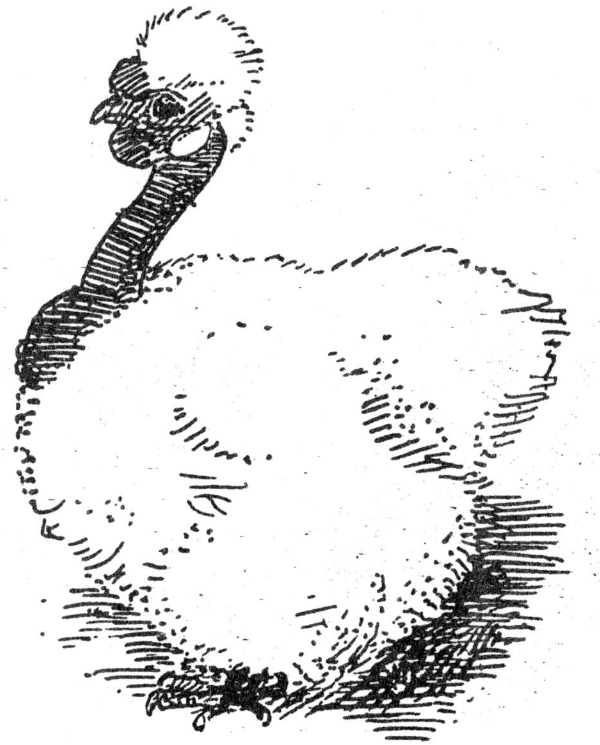

A Cross from A Silkie and Rhode Island Red.

A COMPLEX BREED

The Silkie fowl is a complex bird, possessing many unusual features and therefore genes which produce these characteristics. Examples are as follows:

Silkie feathers	**Single autosomal gene; recessive**
Comb	**Rose comb dominant; Silkie is between walnut and rose. A cross between pea and rose comb produce walnut - possibly the basis of the Silkie 'Cushion' comb. If crossed Walnut X Walnut the F2 result would follow the normal pattern of ratio: 9 walnut; 3 pea; 3 rose and 1 single.**
Polydactyl Five toes	**Autosomal gene; incomplete dominance.**
Crest, Muff & Beard	**Ditto, but separate genes for (a) crest, and (b) muff and beard.**

In addition, there are feathered legs, black skin and bones, and the dark, mulberry face, turquoise ear lobes, and the cushion-type comb. Not enough research appears to have been carried out on these features to allow the separate genes and characterisics to be stated with certainty.

Note: *Incomplete dominance* means that the result is not the normal 3:1; compare the crossing of Andalusian Blues: Blue X Blue gives a mixture of black, blue, and black (or blue) and white splash.(1:2:1).; ie, 50% Blues and the balance a mixture.

BIBLIOGRAPHY

Numerous magazines and books were used for the research; many, especially the *Feathered World Year Books*, are long out of print. Titles of significance are:

Bantams, Entwisle W F , Elsted, England, 1996 (reprint of 1894 edition)

Bantam Breeding & Genetics, Jeffry F P., Hindhead, 1977.

Silkies, Jeffrey, F P, American Bantam Assoc,

Genetics & Evolution of the Domestic Fowl, Stevens L , Cambridge, 1991

Natural Incubation & Rearing, Batty, J. Elsted, 1995.

Poultry Diseases Under Modern Management, Coutts, G., Elsted, 1996.

Poultry Breeding, Hagedoorn A L and Sykes G., London, 1953

Mendelism, Punnett, R C , London, 1911

Our Poultry, Harrison Weir, London, 1902

Illustrated Book of Poultry, Wright L., London, many dates c 1895 to 1930s.

Modern Poultry Keeping, Scott G R, London, 1928.

Poultry Breeding & Management, Broomhead Wm W., London, c. 1930.

Races of Domestic Poultry, Brown Edward, Elsted, reprint of 1929 edition.

INDEX